D0758541

The Story of the Solar System

The bodies of our Solar System have orbited continuously around the Sun since their formation. But they have not always been there, and conditions have not always been as they are today. *The Story of the Solar System* explains how our Solar System came into existence, how it has evolved and how it might end billions of years from now. After a brief historical introduction to theories of the formation and structure of the Solar System, the book illustrates the birth of the Sun, and then explains the steps that built up the bodies of the Solar System. With the use of vivid illustrations, the planets, moons, asteroids and comets are described in detail – when and how they were made, what they are made of, and what they look like. Comparison of these objects, and analysis of how they have changed and evolved since birth, is followed by a look towards the end of the Solar System's existence and beyond. Fully illustrated with beautiful, astronomically accurate paintings, this book will fascinate anyone with an interest in our Solar System.

MARK A. GARLICK obtained his PhD in astrophysics from the Mullard Space Science Laboratory in Surrey, England. He is a member of the International Association of Astronomical Artists, and currently works as a freelance science writer and astronomical illustrator.

The Story of the Solar System

Written and illustrated by

Mark A. Garlick

CAMBRIDGE
UNIVERSITY PRESS

PUBLISHED BY THE PRESS SYNDICATE OF THE UNIVERSITY OF CAMBRIDGE
The Pitt Building, Trumpington Street, Cambridge, United Kingdom

CAMBRIDGE UNIVERSITY PRESS
The Edinburgh Building, Cambridge CB2 2RU, UK
40 West 20th Street, New York, NY 10011–4211, USA
477 Williamstown Road, Port Melbourne, VIC 3207, Australia
Ruiz de Alarcón 13, 28014 Madrid, Spain
Dock House, The Waterfront, Cape Town 8001, South Africa

http://www.cambridge.org

First published 2002

Printed in the United Kingdom at the University Press, Cambridge

Typeface Trump Mediaeval 10/13.5 pt. System QuarkXPress™

A catalogue record for this book is available from the British Library

Library of Congress Cataloguing in Publication data

Garlick, Mark A. (Mark Antony), 1968–
Story of the solar system / written and illustrated by Mark A. Garlick.
p. cm.
Includes bibliographical references and index.
ISBN 0 521 80336 5
1. Solar system. I. Title.

QB501.G37 2002
523.2¢09–dc21 2001035680

ISBN 0 521 80336 5 hardback

Contents

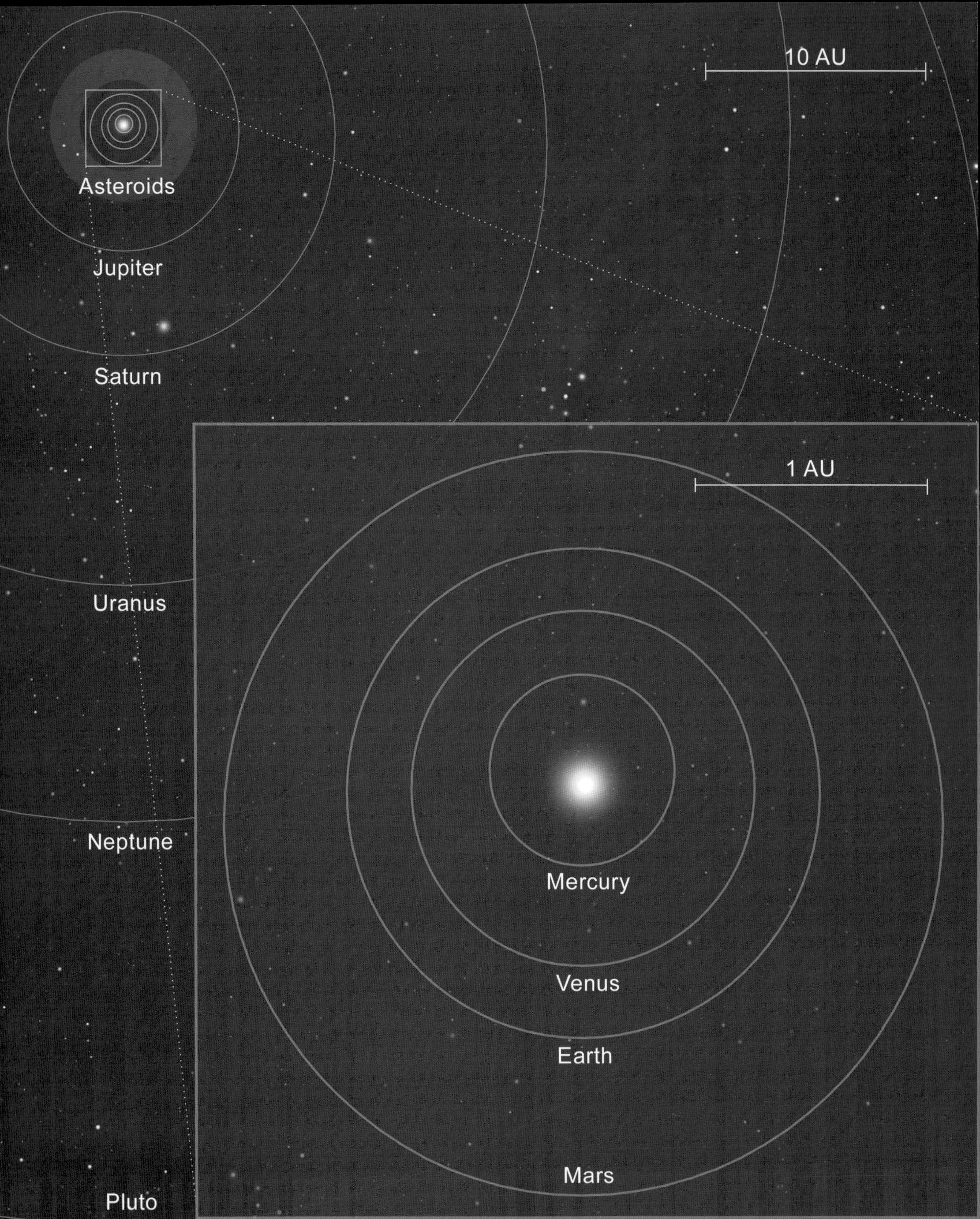
10 AU
Asteroids
Jupiter
Saturn
Uranus
Neptune
Pluto
1 AU
Mercury
Venus
Earth
Mars

Introduction

The Sun, its nine planets and their satellites, the asteroids and the comets – together, these are the elements that comprise the Solar System. In this book we shall meet them in detail. We shall come to know their properties, their place in the Solar System, what they look like and how they compare with one another. We will learn what they are made of, when and how they were made. We will discover what the Solar System's various contents have endured since their fiery birth. And, lastly, we shall see what will happen to them – to the Solar System as a whole – in the far, distant future, billions of years from now, as the tired star we call the Sun passes into old age, and beyond. These and other issues are all part of a great story – the story of the Solar System.

Image opposite: A schematic representation of the planets in their orbits around the Sun, shown to scale. Most of the orbits are near circles, in the same plane – called the ecliptic – but Mercury, Mars and especially Pluto have elliptical orbits with the Sun off-centre, at one focus. Note the order-of-magnitude difference in scale between the zones of the inner and the outer planets – the inner zone is enlarged at bottom right.

Overview of the Solar System

What is the shape of the Solar System? Where are the various objects within it to be found, and how do they move in relation to each other? These are important questions. For, unless we can answer them as accurately as possible, we shall be doomed to failure in our treatment of an even more fundamental issue, dealt with in detail in this book: the origin of the Solar System. So perhaps it would be prudent to spend a little time putting together what we currently know about the Solar System of which we are all a part.

The first thing to establish is that the centre of our planetary system is solar territory. It is the residence of the yellow star that we call the Sun – not the Earth or any of the other major bodies that comprise the Solar System. This may sound like a monumentally naïve statement, but think again. The concept of a Sun-centred, or heliocentric, Solar System was laughed at – even considered fiercely heretical in the Western world – until less than 400 years ago. Before that the generally accepted view was that the Earth lay at the centre, and that the Sun, the Moon and the other known planets (then five) went around it. This was the model that the Egyptian scientist Claudius Ptolemaeus (Ptolemy) propounded in the second century AD. It wasn't until 1543 that the Polish astronomer and churchman Nicolaus Copernicus (1473–1543) published the theory that dared to displace the Earth from the centre of it all and put the Sun in its place. Not surprisingly, Copernicus' theory faced extreme religious opposition. Indeed, Copernicus had the foresight to see how his work would be viewed and, not wishing to confront charges of heresy, held back the publication until the year of his death. In any case the Copernican theory was not perfect either. While it was revolutionary in putting the Sun in the middle, the planetary orbits were wrong. Decades later, it was the German astronomer Johannes Kepler (1571–1630) who found the correct answer. The planets do not quite move in circular orbits. Rather, their orbits are very slightly

Image above: When shown on the same scale, the planets are seen to bunch into three broad types. Those closest to the Sun (bottom) are small and rocky and are known as the terrestrial planets. Jupiter and Saturn are 11.2 and 9.5 times larger than the Earth respectively and are known as gas giants. Uranus and Neptune are intermediate in size and are known as ice giants. Tiny Pluto and its moon Charon do not fit any of these classes and are often considered to belong to the so-called Kuiper-belt objects – icy and rocky bodies orbiting beyond Neptune. Even the largest world, Jupiter, is still only one-tenth the size of the Sun.

elliptical – a path that looks a bit like a squashed circle. Along with Italian observer Galileo Galilei (1564–1642), Kepler was instrumental in confirming once and for all that the Ptolemaic view was dead wrong – despite its having held sway for an astonishing 1500 years.

Since then our understanding of the Solar System has undergone refinements. Of course, more and more discoveries are being made all the time. But here is a summary of some of the Solar System's major characteristics known to date.

1. The Sun is at the centre.
2. All nine planets move around the Sun counter-clockwise as seen from 'above'.
3. Their orbits are truly elliptical but most are nearly circular.
4. Most planetary orbits are within a few degrees of the same plane, the ecliptic.
5. All but three of the planets spin counter-clockwise as seen from 'above'.
6. Most planetary satellites have the same orbital and spin directions as the planets.
7. The four planets closest to the Sun – the terrestrials – are rocky and metallic.

8. The next four planets out from the Sun – the giants – are made of hydrogen and helium.
9. The giants and their orbits are ten times larger than the sizes and orbits of the terrestrials.
10. The last planet, Pluto, is an oddball, fitting none of the above classes.

Thus, the picture that emerges is that of an orderly Solar System, with everything moving and spinning in the same direction and in almost the same plane. Pluto is the only planet whose orbit is sharply inclined to the ecliptic, at more than 17 degrees. Apart from this world, the Solar System is flatter, relatively speaking, than a dinner plate. It is shaped like a disc.

These properties aside, our Solar System has several other important characteristics. We must remember that the Earth shares its home not only with eight other planets, but also a whole multitude of smaller bits and pieces known as asteroids and comets. The asteroids, irregularly shaped chunks of metal and rock, are found mainly between the orbits of the terrestrials and the giants, and again occupy a broadly disc-like environment known as the asteroid belt. The comets, small icy bodies, have two homes. Some lurk beyond the giants in a disc called the Kuiper belt, and trillions more exist a thousand times further from the Sun than Pluto. They surround our star in a vast spherical structure known as the Oort cloud. This, then, is the true extent of the Solar System.

Theories for the Origin of the Solar System

But where did the bodies of the Solar System come from? It's a question that has been puzzled over for thousands of years. The earliest explanations were myths and legends, or irrational tales that stemmed from religious arguments. Indeed, it was only as recently as a few centuries ago that scientists and philosophers, looking at how the Sun, the Earth and the other planets actually behaved, how they moved, started to put forward the first scientific theories to explain the origin of the Sun and its small family. Of course, many of the Solar System's known characteristics as outlined above are recent discoveries. The Kuiper belt and the Oort cloud, for example, were first identified in the mid-twentieth century. So it is not surprising that the earliest attempts to understand the formation of the Solar System were flawed. For they were formulated at times when we had yet to acquire the full picture. This is not to say that we have the complete picture right now. But we certainly have a fuller one – and our improved knowledge of physics helps in our quest for the truth too.

One of the first people to formulate an origin for the Solar System in a scientific way was the French philosopher and mathematician René Descartes (1596–1650). Descartes lived in a time that predated Sir Isaac Newton (1642–1727) – before, therefore, the concept of gravity. Thus, Descartes' personal view was that matter did not move of its own accord, but did so under the influence of God. He imagined that the Universe was filled with vortices of swirling particles, and in 1644 suggested that the Sun and the planets condensed from a particularly large vortex that had somehow contracted. His theory explained the broadly circular motions of the planets, and interestingly he was on the right general track with his idea of contraction. But, we know now that matter does not behave the way he thought it did, and Descartes' theory does not fit the data.

Then, in 1745, another Frenchman put forward an alternative idea. His name was Georges-Louis Leclerc, comte de Buffon (1707–1788). Buffon suggested that a large comet passed close to the Sun and pulled a great arc of solar material out into space, from which the planets later condensed. He did not attempt to explain where the Sun had come from. Interestingly enough, this mechanism – the 'encounter theory' – was revisited in 1900 when two astronomers suggested that the Sun's encounter had been not with a large comet, but with a passing star. But both ideas are wrong. The material drawn from the Sun would have been too hot to form planets. And on average the stars are separated like cherries spaced miles apart – the chances of any star coming remotely close to another, even over the age of our Milky Way galaxy, are very small indeed. If correct, the more recent of the two encounter theories would have us believe that our Solar System is a rarity, the happy outcome of a sheer coincidence, and thus one of just a handful in the galaxy of 200 billion stars to which our Sun belongs. But as we shall see below, planetary systems are the norm, not the exception. Again, this theory does not fit the data.

The theory most broadly correct – or at least currently accepted – for the origin of the Solar System was first formulated in 1755 by the German philosopher Immanuel Kant (1724–1804). Kant believed that the Sun and the planets condensed from a gargantuan disc of gas and dust that had evolved from a cloud of interstellar material. However, his theory went relatively unnoticed, and it wasn't until Pierre-Simon, marquis de Laplace (1749–1827) independently came up with the same idea 54 years later that the model garnered attention. Kant and Laplace succeeded where Descartes had failed because their work included the Newtonian concept of gravity. Their view was that a collapsing interstellar cloud would flatten out by virtue of its rotation. The Sun would emerge in the centre, while the planets would form further out in the disc, condensing from concentric rings of material shed by the central star. This became known as the 'nebular hypothesis'.

The advantages of the nebular hypothesis are many. It produces a discal, heliocentric Solar System with planets in neat, near-circular orbits, all orbiting and spinning in the same direction – satisfying characteristics 1–6 above. But there was one big problem with the idea: it left the Sun spinning much too quickly. The Sun, which rotates on its axis just as the planets do, spins once in about 30 days. (It actually rotates at different speeds depending on solar latitude.) But according to the nebular hypothesis it ought to be

Image opposite: The modern theory for the origin of the Solar System is based on models proposed in the eighteenth century by Kant and Laplace. Known as the nebular hypothesis, it proposes that the Sun, the planets, the asteroids and the comets all formed at the same time when a cloud of interstellar material collapsed under gravity and flattened out because of rotation. The Sun formed at the centre, and the planets gradually accreted in the disc.

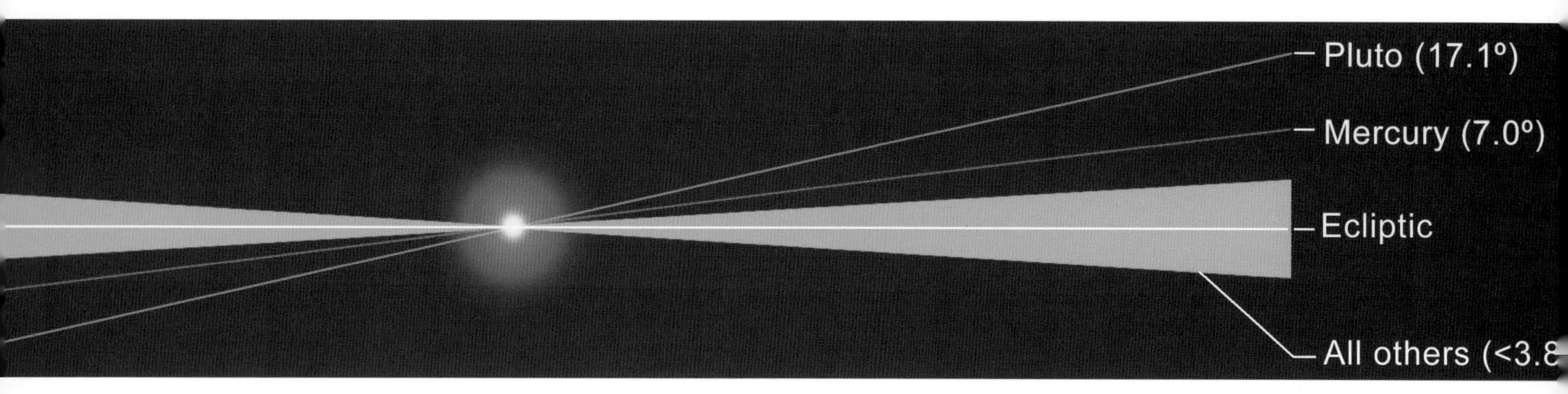

Image above: With the exception of Pluto and Mercury, all the planets orbit the Sun very close to the ecliptic, defined as the plane in which Earth orbits. Seen from the side therefore, most of the planets reside in a thin disc, here represented as a pair of orange triangles.

spinning almost 400 times faster. In scientific parlance, the Sun has very little of its original angular momentum left, and this is known as the angular momentum problem. Still, modern astronomers have not discarded the nebular theory. Indeed, they have adapted it and refined it to the point where it now produces a more slowly rotating Sun and satisfies points 7–10. More importantly, as observational technology has improved, it has emerged that the Milky Way galaxy is full of exactly the kind of object that, according to Kant and Laplace, built our Solar System: vast pancakes of warm gas and dust known as protoplanetary discs. Nowadays, the one that spawned our own Solar System is referred to as the Solar Nebula.

Still, even the Solar Nebula model has problems. Astronomers are not only finding protoplanetary discs; they are also chalking up new planets beyond our Solar System – so-called exoplanets or extrasolar planets, surrounding other stars – and they are doing so at an alarming rate. Already, in just five years, the number of known planetary systems has climbed from zero to dozens. The trouble for the nebular model is that, although it accounts for many of the properties of the Solar System, it fails to reproduce the detailed characteristics of many of these new systems. Some of them, for example, have very massive planets in extremely elliptical orbits, not the near-circular orbits most solar planets have. Other stars have massive planets very, very close to their central stars, often with orbital periods – 'years' – of just a few Earth days! These massive planets are probably gaseous, like Jupiter and Saturn. Yet there is no easy way to see how they could have formed so close to their parent stars. Giant planets are generally believed to have formed where they did in our Solar System, far from the Sun, because it was only at these distances that the temperatures dropped to the point that ices could condense. Closer in, it was much too hot, and only small planets of rock and iron could grow.

The bottom line is that there is still a long way to go before we truly have a model that can faithfully reproduce the observed properties of every known planetary system, including ours. Indeed, it is likely that no model will ever be found. In our own Solar System, for example, many of the planets have the properties they do because of unpredictable cosmic impacts long ago in their past. If the Solar System formed all over again, the Earth might not have its Moon, and Pluto could well have a more normal, near-circular orbit – these are just two of many of the Solar System's properties that might have been very different had things not gone the way they had. Still, the general picture of stars and planets forming from rotating discs seems well established. More than any other theory, the nebular hypothesis is the one that fits the data. This is the model that I assume in this book.

Story of the Solar System

But this book is not just about the Solar System's origins. Indeed, this is only part of the story of the Solar System, covered comprehensively in step-by-step fashion in Parts 1 and 2. Part 3 also touches on this issue, but is largely concerned with presenting a detailed inventory and cross-comparison of the Solar System's contents, and an analysis of how they have changed and evolved since birth. Lastly, Part 4 looks to the future. It deals with our planetary system's eventual demise, in a time far too distant for us truly to comprehend.

A look to the future may sound somewhat bold. Certainly we shall not be around to see what will happen to our Sun, the Earth and all the rest of it even deeper into the future than we can trace their origins into the past. How will we ever know for sure if our theories are correct? We almost certainly will not. But we can make good guesses by observation and data acquisition. Astronomers have studied enough stars now to have a good understanding not only of what the Sun has gone through already, since birth, but of what lies ahead in the next several billion years that will lead ultimately to its downfall. A good way to understand how astronomers know this is to imagine photographs in a family album. Individually the pictures tell very little about the human life cycle. But by studying images of people at various stages through their lives, it is possible to deduce how humans change physically with time. They start off small, grow steadily taller, reach a sort of plateau, grow wrinkled and bent – those that don't age gracefully! – and then cease to exist. It's the same with the stars. There are so many of them, each at different stages in their evolution, that taken together they tell a story – the story of the life of a single, general star, from the cradle to the grave.

And so it is by theorising, and by checking theories with observations, that astronomers have reached their current understanding of the Solar System, past, present and even future. Now, let's have a look at that great story in detail, starting, where most tales do, at the very beginning.

Part 1

Genesis of the Sun and Solar Nebula

'Let there be light'

Genesis 1:3

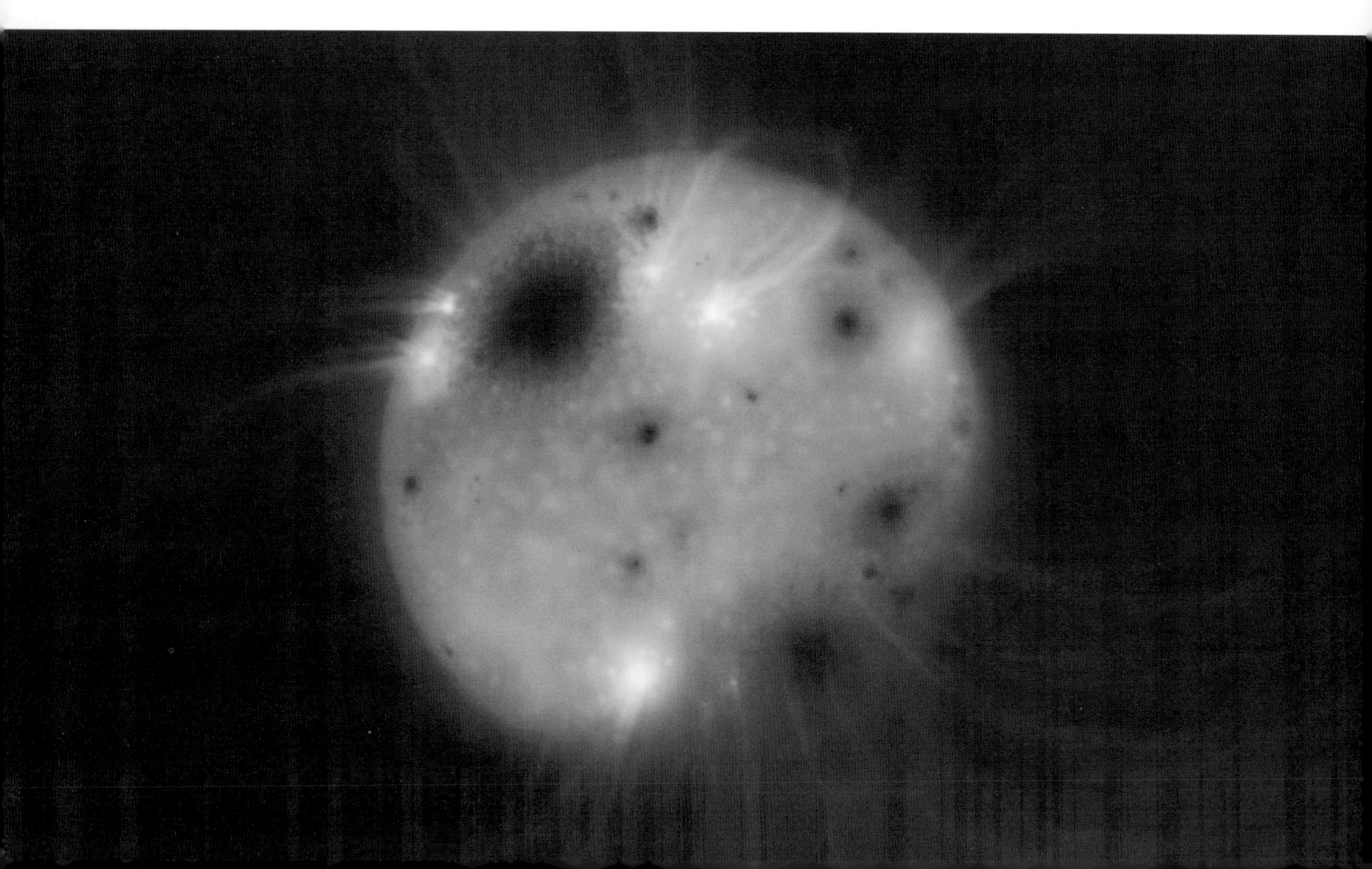

Thirty million to 50 million years. That's all the time it took to form the star we call the Sun. This may sound like a long time, but let's put it in perspective. Since the last dinosaurs walked the planet, enough time has passed for at least one and possibly two stars like the Sun to have formed, one after the other – utterly from scratch. The details of this miraculous creation are not exceptionally well understood, but astronomers at least have a good grounding in the basics. Perhaps ironically, one star's birth starts at the other end of the line – when other stars die.

Generally speaking, stars make their exit in one of two ways. A low-mass star like the Sun eventually expands its outermost layers until the star becomes a gross, bloated caricature of itself: a red giant. Gradually, the star's envelope expands outwards, all the time becoming thinner, until the dense core of the star is revealed. Such an object is known as a white dwarf. It is a tiny and, at first, white-hot object with a stellar mass – yet confined to live out the rest of its existence within the limits of a planet's radius. The rest of the star meanwhile, the cast-off atmosphere, grows larger and larger. Eventually it becomes nothing but a thin fog of gas spread over more than a light-year. This is the fate that awaits our Sun, as we shall see in detail in Part 4. By contrast, a heavier star dies much more spectacularly. It blows itself to smithereens in a star-shattering explosion called a supernova. The star's gases are jettisoned into space where, again, they disperse. Whichever way a star finally meets its doom, much of its material has the same ultimate destiny: it is flung back into the galaxy. Over billions of years, these stellar remains accumulate and assemble themselves into the enormous clouds that astronomers refer to collectively as interstellar matter.

But that is not the end of the story. In fact, it is our starting point. For the Universe is the ultimate recycling machine. Starting around 4660 million years ago, from the ashes of dead stars, a new one eventually grew: a star known as the Sun.

Time zero
Giant Molecular Cloud

Image above: Sometimes, newly forming stars within molecular clouds energise the gases and make them shine. This is why the Orion Nebula, 1500 light-years away, is so conspicuous. *Courtesy C. R. O'Dell and NASA.*

Image opposite: A supernova, the cataclysmic explosion of a dying star, drives shockwaves into a nearby molecular cloud and rips it to pieces. These fragments will later begin to collapse under their own gravity, and one of them is destined to become the Sun.

Before 4660 million years ago, our Solar System existed as little more than a cloud of raw materials. The Sun, the planets, trees, people, the AIDS virus – all came from this single, rarefied cloud of gas and dust particles. These patches of interstellar fog were as common billions of years ago as they are now. They are known as giant molecular clouds.

Orbiting the nucleus of a galaxy called the Milky Way, about two-thirds of the way out from the centre, this ancient cloud from which the Solar System sprang was about 50–100 light-years across, similar in size to its modern cousins. And again, like today's giant molecular clouds, it presumably contained enough material to outweigh millions of stars like the Sun. Most of its mass, about 73 per cent of it, was made up of molecular hydrogen, a gas in which the hydrogen atoms are glued together in twos to make simple molecules. The rest of the cloud's material was in the form of helium, with traces of heavier elements such as carbon, nitrogen and oxygen, and particles of silicate materials – fragments that astronomers like to lump under the category of 'dust'. With between a few thousand and a million gas molecules per cubic centimetre, the cloud would have been recognised as better than a first-class vacuum by today's standards. And it was very cold, around −250 Celsius, barely hotter than interstellar space itself. Molecular hydrogen cannot survive at very much higher temperatures, because the energy shakes the molecules apart. So the cold kept the molecules intact. But the cloud was nevertheless in danger of destruction.

A molecular cloud is like an interstellar house of cards, forever on the verge of disintegration. A push, a pull, anything could have triggered this ancient cloud's demise – and there are lots of potential triggers spread over 100 light-years of interstellar space. The cloud might have passed close to a massive star whose gravitational tug stirred up the molecules within the nebula. Or the cloud could perhaps have drifted within close range of a supernova explosion, the shockwaves from the dying star burrowing into the cosmic smog and compressing its gases. It would have taken only one such event to collapse the house of cards, to make the cloud fall in on itself under gravity.

Something like this must have happened to our ancient molecular cloud about 4660 million years ago. It was the first step in the process that would eventually lead to the formation of a certain star.

2 000 000 years
Solar Globule

Image opposite: A globule is a fragment of a molecular cloud, inside of which a star is being made. Because the dust and gas accelerates inwards faster near the centre than further away, the more distant material gets left behind in a shell while a dense core develops further in. The red material is background gas in a more distant, brighter and unrelated nebula.

Once the collapse of the giant molecular cloud had started, it continued under its own momentum. By the time two million years had passed, a multitude of nuclei had developed in the cloud, regions where the density was higher than average. These concentrations began to pull in more gas from their surroundings by virtue of their stronger gravity, and the original cloud fragmented into hundreds or even thousands of small, dense cores. Most of them would later form stars. One of them was destined to become the Sun.

By now, the cloud core from which the Sun would form was perhaps a tenth of a light-year across, more than a hundred times the present size of the Solar System out to Pluto. Gradually, this tight clump of gas continued to fall in on itself like a slow-motion demolished chimney stack, a process known as gravitational freefall. The innermost regions fell the fastest; they were closest to the central condensation where the gravitational pull was greatest. The outermost edges of the cloud core took longer to succumb to their inevitable fall. Thus, because of these differences in infall rates, the cloud's contraction essentially amounted to an implosion, an explosion in reverse. In time, as the gas closest to the centre plunged inward and accelerated, the material there grew steadily hotter, the atoms and molecules within it rubbing against each other frantically. After perhaps millions of years in a deep freeze, the molecular cloud was finally warming up. The eventual result was a gas and dust cocoon: a shell of dark material surrounding a denser, warmer core. Such an object is known as a globule. It was the Sun's incubator.

As with all globules, the solar globule was dark. It emitted no light at all. But a bit later in its evolution, as it gradually warmed, it was a strong emitter of heat radiation or infrared. Only an infrared telescope, and possibly a radio telescope, would have been able to penetrate the gas and dust and home in on the low-energy radiation coming from the globule's gently warming core, and see the first, feeble stirrings of the yellow star that the globule would one day become.

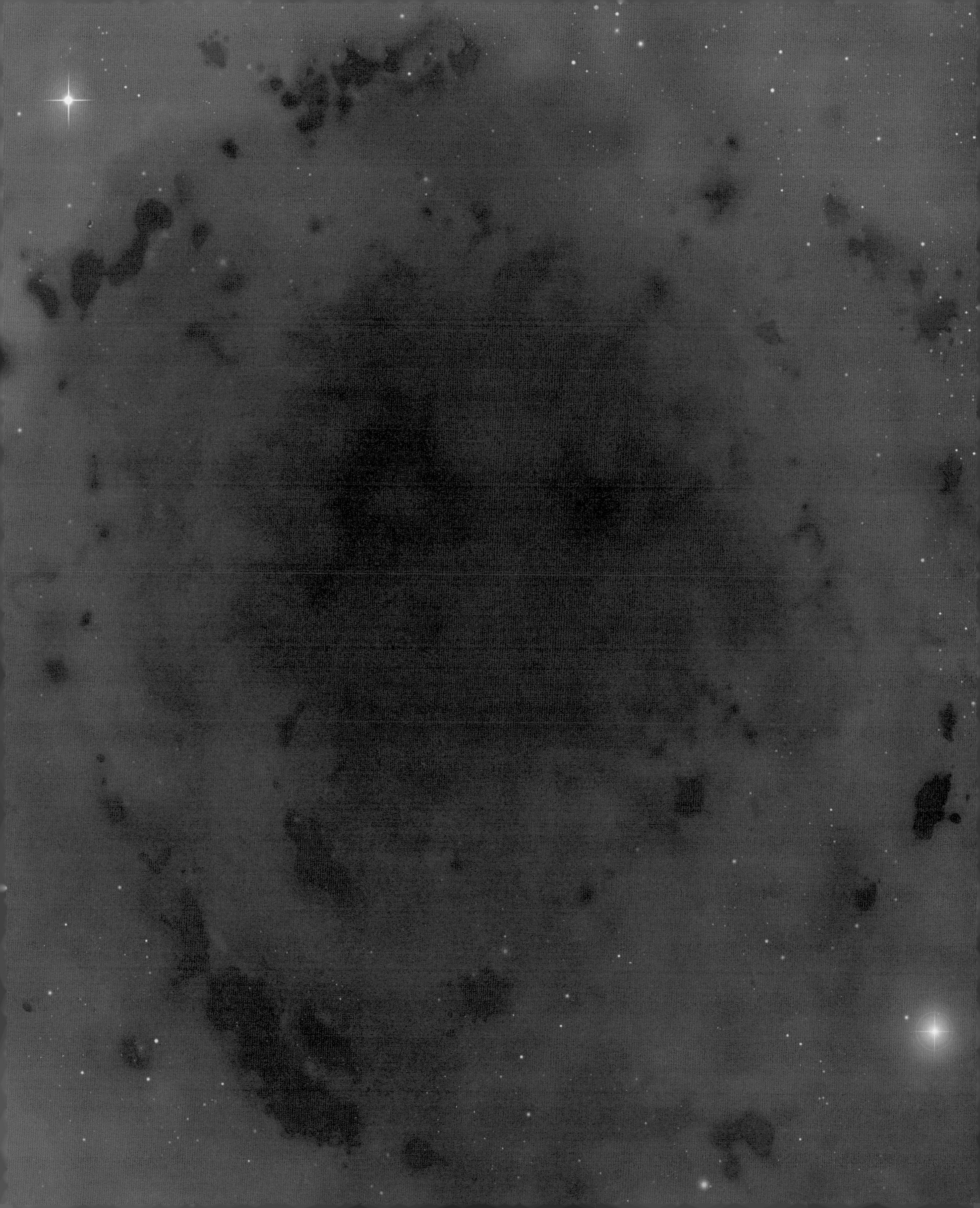

2 030 000 years
Protosun

Image opposite: The protosun as it might have appeared billions of years ago – if we had been able to peer inside the thick cocoon of gas and dust that still encased it. The surface in this depiction, which shows the protosun at an advanced stage, is now hot enough to glow, its temperature around a few thousand degrees.

Over tens of thousands of years, the gases inside the globule continued to fall away from the inside edge of the cocoon, pulled inexorably towards that dense core at the centre. By now, the core of the globule was taking on a definite shape – a gargantuan ball, about the size of the present-day Solar System out to Pluto. Its surface was still too cold to glow optically. But, at last, its central regions had warmed up significantly – to about 10 000 Celsius – and the molecules there had split into atoms of hydrogen.

This marked an important point in the development of the Sun. At this temperature, the cloud core was now hot enough for the radiation it emitted to carry a significant punch. Radiation is composed of tiny packets of energy called photons, each of which can be likened to a subatomic particle. If there are enough of these photons emitted every second they can hit like a hail of bullets, a barrage of electromagnetic force known as radiation pressure. Before this point the core of the globule had been emitting too few photons to exert a noticeable force. Now, though, as the growing waves of radiation streamed away from the warming core they slammed into the outermost regions of the globule where the gases were less dense, and slightly hindered their inbound journey. Thus the contraction of the core slowed, but it did not stop, so overwhelming was the inward pull of gravity. The very centre of the core was also dense enough now that it was beginning to become opaque to the heat radiation generated inside it. The energy could no longer escape so easily, so from here on the nucleus heated up much faster as it shrank. The build-up of heat thus slowed the contraction ever more, and the core grew at a much slower pace. It had reached a configuration that astronomers ennoble with the term 'protostar'.

By this time, the protostar – 'protosun' in this case – had developed a marked rotation. Just as water being sucked down a plug hole spirals around before it falls in, so the gases that had fallen into the protosun had begun to swirl about. And in the same way that a yo-yo spun around on its string spins faster as the string winds around a finger – owing to a concept known as the conservation of angular momentum – so the infalling gases had increased their angular speed as their long journey inwards had progressed. As the protosun grew smaller and hotter, therefore, it began to spin faster and faster.

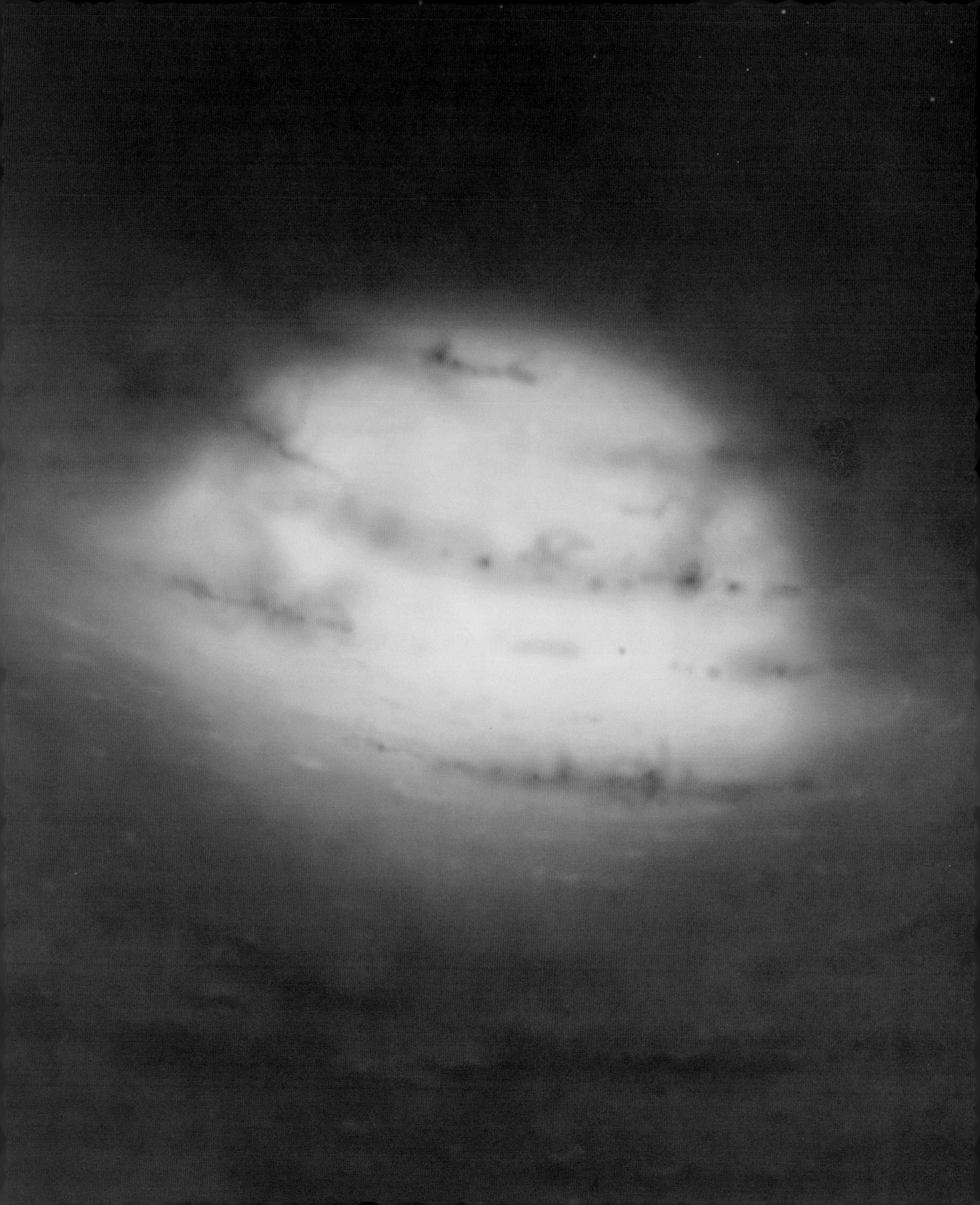

2 130 000 years
Solar Nebula

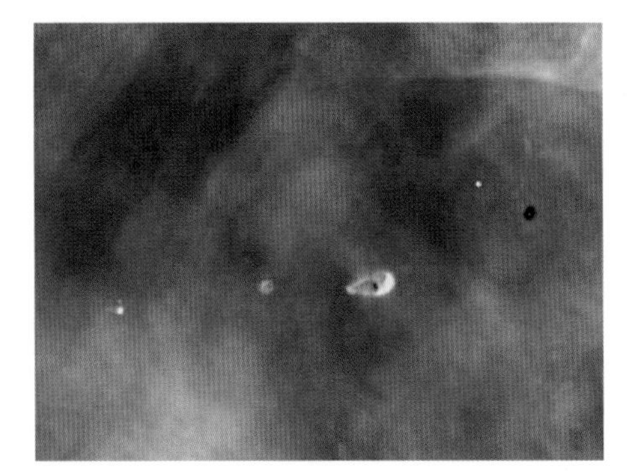

Image above: Protoplanetary discs, or proplyds, are common in the Milky Way, direct proof that this is how planetary systems form. The proplyds in the Orion Nebula are perhaps the most striking examples, as this Hubble Space Telescope image shows. *Courtesy C. R. O'Dell and NASA.*

Image opposite: The Solar Nebula, a swirling pancake of gas and dust, surrounds the newly forming star known as the Sun. Later, planets will form there.

The protosun's collapse continued. Within 100 000 years or so it had become a swollen semi-spherical mass, flattened at the poles by rotation. Its surface temperature of the order of a few thousand degrees, the protosun was at last glowing visibly for the first time. And its diameter was by now roughly equal to that of the present orbit of Mercury – about 100 million kilometres. But the newly forming star was no longer alone. Over the aeons the rapid rotation of the infalling matter had flattened out the gases like a pizza dough spun in the air. Now, a huge pancake of turbulent, swirling gas and dust surrounded the protosun right down to its surface. Thinner near the centre, flared vertically at the edges, this structure is known as the Solar Nebula.

The Solar Nebula measured about 100 to 200 astronomical units (AU) across, where 1 AU is defined as the current distance from the Earth to the Sun, 150 million kilometres. The disc would have contained about 1–10 per cent of the current mass of the Sun – most of it in the form of gas, with about 0.1 per cent of a solar mass locked up inside particles of dust. Near the centre of the disc, close to the seething protosun, the temperature may have exceeded 2000 Celsius. Here, where things were hot and important, the disc may have been hot enough to emit its own visible radiation – in any case it would have shined optically by virtue of the light it reflected from the protosun. Further out in the disc the temperature dropped rapidly with distance, though, and it would have shone only in the infrared. At about 5 AU, the current location of the planet Jupiter, the temperature dipped below −70 Celsius. And on the outside edges, where the material was more rarefied and the disc vertically flared, it was even colder. This vast reservoir of material was the raw substance out of which the planets would soon begin to condense, as will become evident in Part 2. It is thus known as a protoplanetary disc, or proplyd for short.

By now, much of the original globule had been consumed. Most of it had fallen into the protosun, and the rest into the disc. At last, with the globule eaten away, the newly forming star was revealed to the exterior cosmos for the first time, as it prepared itself for the next – and most violent – stage in its formation: the T-Tauri phase.

3 million years
T-Tauri Phase

Image opposite: The Sun during its early T-Tauri phase is still surrounded by a gigantic disc, but the disc's central regions are now swept clear by the whirling magnetic field. Like beads on a wire, blobs of gas leap across this clearing from the disc to the Sun, and fierce flares erupt where the gas strikes the star's toiling surface.

By 3 million years or thereabouts – about 1 million years after the initial collapse of the globule – the protosun had shrunk to a few solar radii. Its temperature at the centre was now around 5 million degrees Celsius, while the surface seethed and bubbled at around 4500 Celsius. At last the object had crossed the line that separates protostars from true stellar objects. It joined the ranks as what astronomers call a T-Tauri star.

Named after a prototypical young stellar object in the constellation Taurus, the T-Tauri phase is one of extreme fury. And as with all T-Tauri stars, this earliest form of solar activity would have been driven – at least in part – by a powerful magnetic field. Because the gases inside the young star were by now fully ionised – a soup of positively and negatively charged elements – their movement as the star rotated effectively amounted to a series of gigantic electric currents. Thus the spinning star developed a global magnetic field in the same way that a wire carrying an electric current does – just as the Sun generates its field even today. During the Sun's T-Tauri phase, though, the star would have been spinning very quickly – once in 8 days compared with once in 30 days – spun up by the swirling gases that had ploughed into it earlier. This means that the T-Tauri Sun's magnetic field was much mightier than at present, and this is what made this phase in the Sun's formation so violent. The Sun was still surrounded by its protoplanetary disc. So, as the Sun whirled around, it dragged its magnetic field through this disc. Where the field and disc connected, vast globs of gas were wrenched out of the surface of the disc and sucked along the field lines, right into the young Sun. And where these packets of gas hit, the troubled star responded with the violent flares that are the hallmarks of the T-Tauri phase of star formation.

Thus the adolescent Sun was very much more violent than the star we know today. It looked the part too. Its larger, cooler surface meant it glowed an angry red, not a soft yellow. And the sunspots that dotted the solar surface then were very much larger than their modern counterparts. Sunspots are generated when the Sun's rotation tangles its magnetic field and creates localised regions of enhanced magnetic field strength. Where these entanglements are greatest, the increased magnetism hinders the flow of gases on the surface and cools those regions down – and they appear as dark patches. Today, the Sun's spots cover less than 1 per cent of its surface. But the T-Tauri Sun would have had sunspot 'continents' covering great stretches of its bloated face.

Perhaps the most awesome aspect of the T-Tauri phase, however, was the molecular outflow. This would come next.

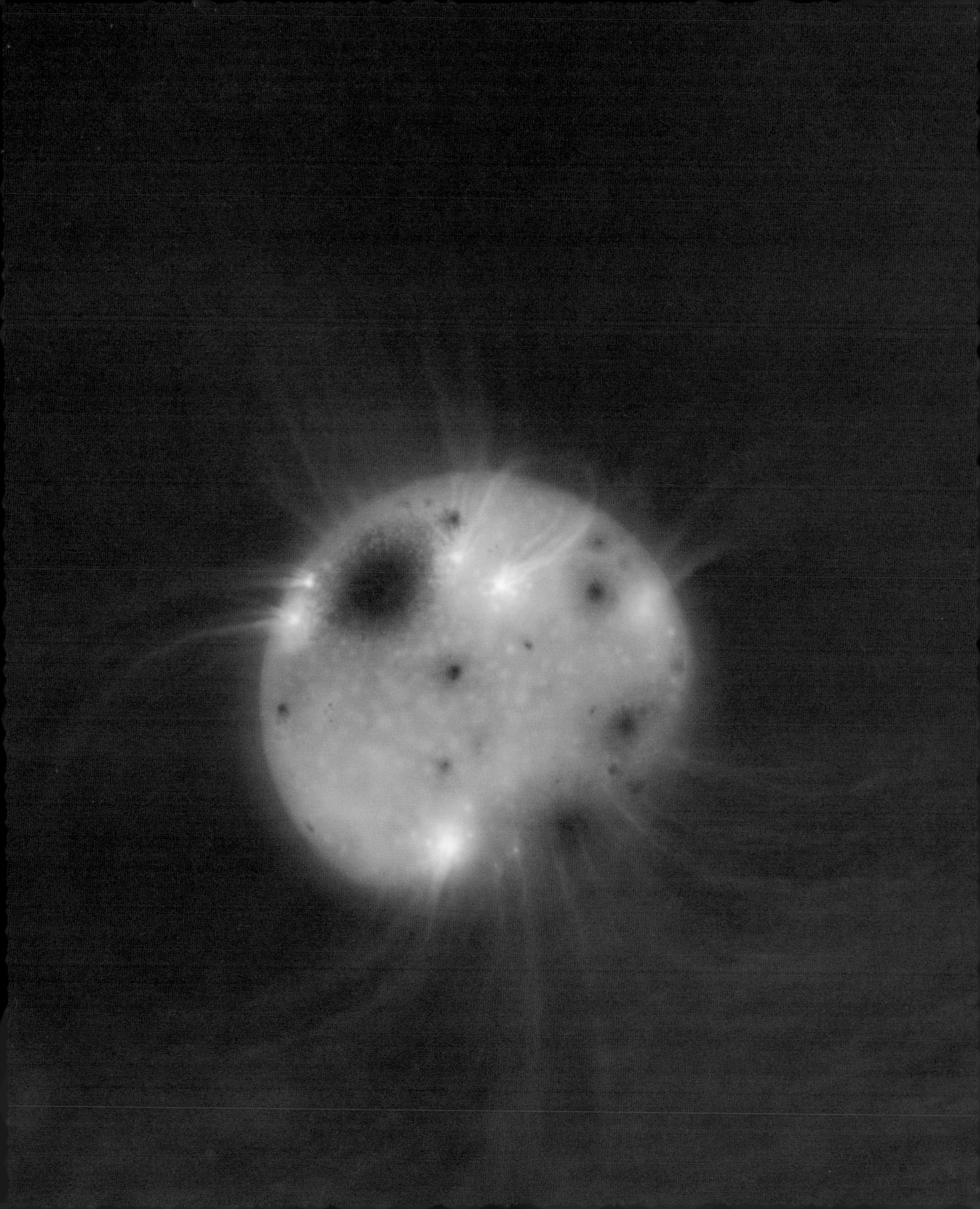

3 million years
Outflow and Post-T-Tauri Phase

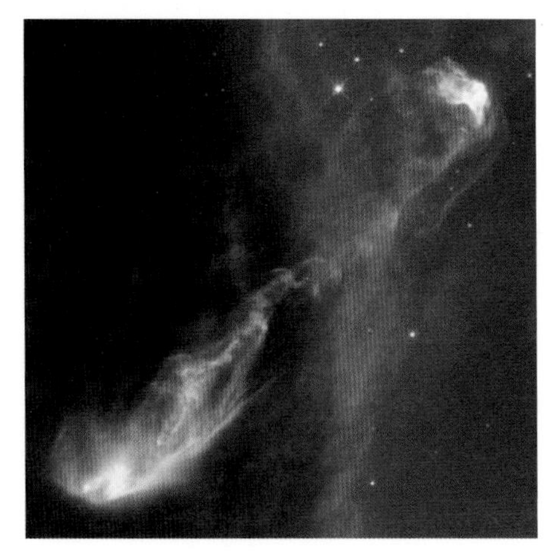

Image opposite: Seen edge-on from a distance of some 20 billion kilometres, the Solar Nebula appears as a bloated, clumpy pancake. Deflected by this disc and focused by magnetic forces, the Sun's T-Tauri wind forms a bipolar outflow: two jets that extend several light-years out into the depths of space.

Image above: Where the outflow crashes into interstellar gas, the energy of the collision makes the gases glow brightly. We see this cosmic pile-up, trillions of kilometres long, as a Herbig–Haro object, one of which is captured in this Hubble Space Telescope image. *Courtesy J. Morse (STScI) and NASA.*

Almost as soon as the Sun hit the T-Tauri stage – possibly even slightly before – it developed what astronomers call a stellar wind. The modern Sun also has one: a sea of charged particles that streams away from the solar surface, out into the depths of the Solar System. But T-Tauri winds are much more furious and contain more mass, moving at speeds of up to 200 kilometres per second.

How T-Tauri winds are generated is still very poorly understood. A possible cause is again the rapid rotation. Some of the gas dragged out of the Solar Nebula disc would have plummeted towards the star's surface. But not all of it. Because the Sun was by now spinning very quickly, some of the gas pulled out of the disc plane was hurled radially outwards, much as water is spun out of the wet clothes in a spin dryer. The result was a steady flow of gas away from the star's surface. However the Sun's T-Tauri wind arose, its effects would have been quite dramatic. As the wind blasted away from the young Sun's surface it banged into the disc and was deflected through a sharp angle out of the disc's orbital plane. The disc might have been threaded by magnetic field lines as well, and these would have channelled the flowing gas further away from the disc and 'up' and 'down' into space. The result was a 'beam' of charged particles blustering away from the young Sun in two opposed directions, perpendicular to the protoplanetary disc. Astronomers call this a bipolar molecular outflow. By now the Sun had stopped amassing material and in fact would lose a significant fraction of its original mass, throughout the life of the wind, via the outflow.

By the time the wind had ceased, a fleeting 10 000 years since it had started, the Sun's mass had begun to stabilise. However, it continued to shrink under gravity because the pressure at its core, though great, was not yet adequate to stop the contraction. All the time the Sun was slowly contracting it was also gradually approaching its modern temperature and luminosity. This was by far the slowest period in the Sun's formation. Even as the Sun's violence ended and it entered the relative calm of the post-T-Tauri phase, a couple of million years after it had started, the Sun still had tens of millions of years to go before reaching full maturity.

30–50 million years
The Main Sequence

Image opposite: An impression of the Sun as we know it, as it has been for the last few billion years. Gone is the angry red colour it had at birth – now the Sun glows a slightly hotter yellow. The sunspots are smaller too, the signs of reduced magnetic activity brought about by a slower rotation.

At last, after a period of perhaps 30 to 50 million years – astronomers still cannot agree on their numbers – the Sun's contraction finally came to an end. Why? Because the Sun's internal temperature had reached an all-time high of 15 000 000 Celsius – and something had begun to happen to its supply of hydrogen.

Hydrogen is the simplest of all elements. Each atom contains just a single subatomic particle called a proton in its nucleus, positively charged. Orbiting this, meanwhile, is a single much smaller particle with exactly the opposite electric charge: an electron. Inside the Sun, these atoms are ionised: the electrons are detached and roam freely in the sea of hydrogen nuclei or protons. Very often, two of these hydrogen nuclei come together. Just as two magnetic poles of like polarity repel each other, so too do two protons. But not if they are brought together with sufficient speed. The speed of particles in a gas can be measured by the gas's temperature. And at 15 000 000 Celsius, the positively charged hydrogen nuclei at the Sun's core were now moving so quickly that when they smashed together they overcame their electrostatic repulsion, and fused as stronger nuclear forces took over. At last, the hydrogen was being consumed, gradually converted into helium in the Sun's core via a chain of nuclear reactions. Energy is a by-product of these reactions. And so the Sun now began to generate a significant amount of power in its core. The pressure of this virgin radiation was so intense that for the first time since the original gas cloud had started to contract, tens of millions of years earlier, the force of gravity had finally met its match. Exactly balanced against further contraction, and slowly metamorphosing hydrogen into helium in its core, the Sun at last got its first taste of the so-called main sequence. It had become a stable star, in a state that astronomers call hydrostatic equilibrium.

This, the ignition of core hydrogen, was the point at which the Sun as we know it was truly born. Called the main sequence, or hydrogen-burning phase, this is by far the longest-lived stage in the life of a star. It took tens of millions of years for the Sun to get to this point – yet so far it has existed for about 100 times longer than that with little change. About 4600 million years later, it is not quite halfway through its main-sequence journey. It still has a long life ahead.

Part 2

Emergence of the Sun's Family

'Space may produce new worlds'

John Milton, *Paradise Lost*

The planets, their moons, the asteroids and the comets – all are part of the Sun's family. And they are just as ancient as their parent. Evidence suggests that the Solar System's contents started to form even while the Sun itself was still only a protostar, almost as soon as the Solar Nebula was in place.

We have seen that, in some ways, the Sun formed in much the same manner in which a sculpture is made. What began as a single, large block of material – the giant molecular cloud – was gradually whittled away to reveal a smaller end product. But the planets' origins are more like those of buildings. They grew bit by bit, from the bottom up, by accumulating steadily larger building blocks. The very first process in the planet-building production line is a familiar concept known as condensation. You can see it in action when somebody wearing spectacles enters a warm room after being outside in the cold. As soon as air-borne water molecules hit the cold lens surfaces, the molecules cool down and stick to the lenses one at a time to produce a thin – and very annoying – film of tiny water droplets. Exactly the same phenomenon was big business in the very earliest stages of the Solar Nebula. As more and more material spiralled from the Solar Nebula into the newly forming Sun, the disc grew less dense. Eventually it became so sparse that its infrared energy could pass through with less hindrance. Thus the heat leaked away into space, the disc began to cool, and its material started to condense – single atoms or molecules grouping together one at a time until they had grown into tiny grains or droplets less than a millionth of a metre across.

But it was not condensation alone that produced the Sun's family. Condensation is only an efficient growth mechanism when the grains or droplets involved are small, because matter is added one atom or molecule at a time. Eventually, as will become clear, the process was replaced by agglomeration and accretion – the building of progressively larger fragments through the accumulation of other fragments, not atoms.

The planet-building processes themselves are reasonably well understood. And yet, even after decades of research, astronomers can agree neither on the timescales involved in the various stages, nor on the sequence in which the events took place. It seems fairly certain that the gas-rich planets Jupiter and possibly Saturn formed very quickly – shortly it will become evident why. The rest, though, is more uncertain. And so what follows represents only one possible sequence in which the various elements of the Sun's family came into being. This, the second part of our story, begins in the Solar Nebula, after the onset of condensation. Time elapsed since the fragmentation and collapse of the giant molecular cloud: 2 200 000 years.

2 200 000 years
Planetesimals and Protoplanets

The Solar Nebula was a rich soup of many different components. Gases such as hydrogen, helium, carbon and oxygen were common. Thus the disc brimmed with molecules – water, ammonia and methane – made from these available gases. Atoms of silicon – the basis of rock – were also abundant, along with metals. But these metals did not exist uniformly throughout the disc. Close to the protosun, where the temperature was around 2000 Celsius, only the very densest materials, such as iron, could condense. So the grains that grew there had a significant iron content. A bit further out, where it was cooler, silicate particles condensed into grains of rock. And at about 5 AU from the centre, the current location of the planet Jupiter, ices began to gather. Here, at what astronomers call the 'snow line', the Solar Nebula was a lot colder – maybe less than −70 Celsius. It was here and beyond where the water, ammonia and methane finally condensed out and froze to form ice crystals.

Thus, with the onset of condensation in the Solar Nebula, the protoplanetary disc soon began to resemble a vast, swirling storm of sand, iron filings and snow, whizzing around the central star at speeds of tens of kilometres per second. Collisions between adjacent particles were of course inevitable. And yet, for the most part, these interactions were fairly gentle, not violent. One way to imagine the scenario is to picture racing cars speeding around their circuit. Naturally the cars travel very fast – relative to the road and the cheering spectators. But, relative to each other, their speeds are much less reduced, hovering around the zero mark. Occasionally one of the cars will nudge up alongside and touch one of the others.

Image below: A glance in the direction of the newly forming Sun (right) from the mid-plane of the Solar Nebula reveals countless particles ranging in sizes from dust grains up to asteroid-like fragments kilometres across. The largest of these are planetesimals, the building blocks of planets.

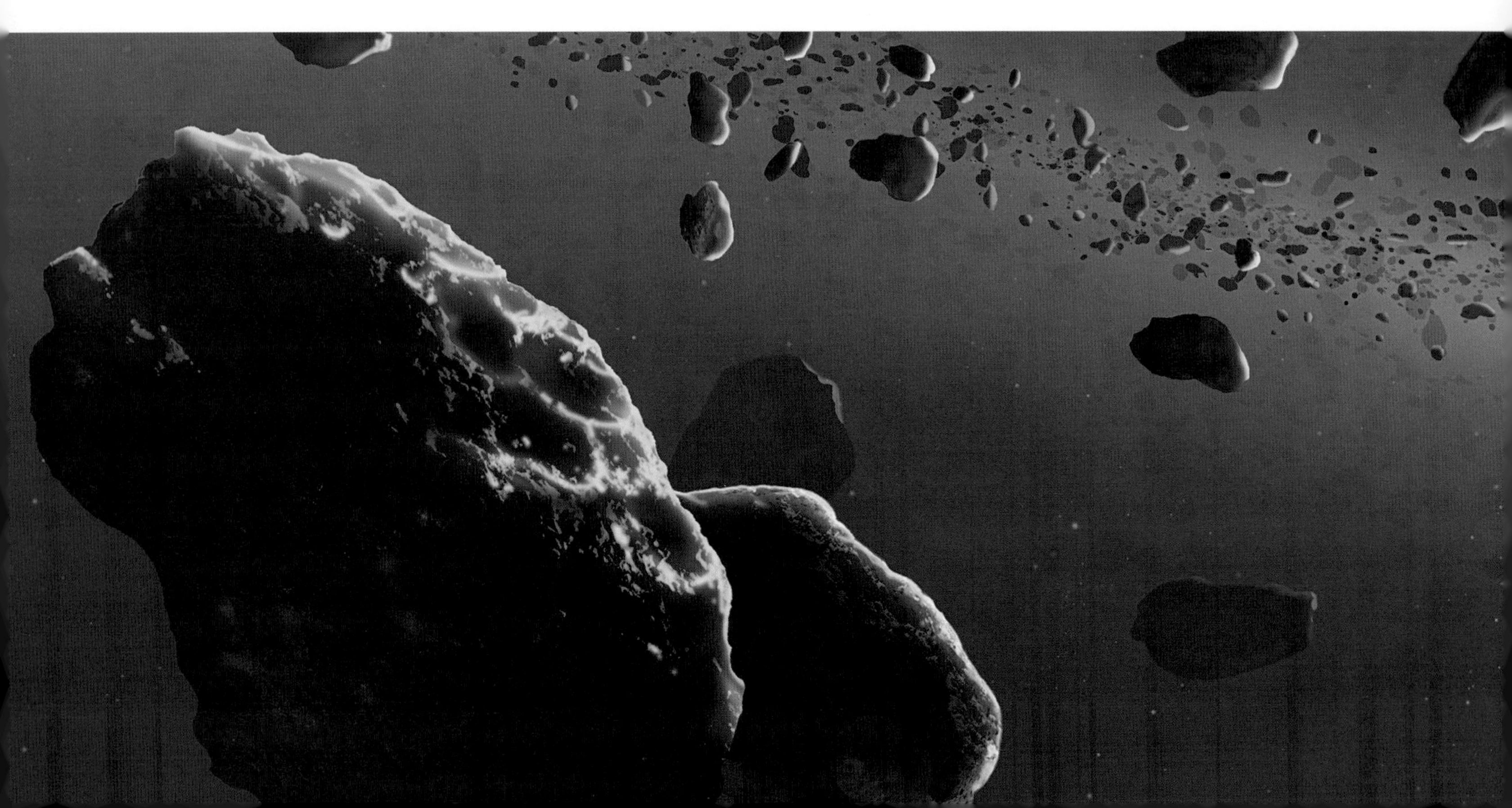

And so it was with the condensed particles in the Solar Nebula. Even though they were moving around so quickly, they were still able to jostle up alongside their neighbours fairly gently. When that happened, many of the particles stuck together, bonded perhaps by electrostatic forces. This is known as agglomeration. Thus, through this process, the first fragments grew steadily larger still. And the results were extremely rapid. Within just a few thousand years of its appearance, the Solar Nebula teamed not only with dust, but also with countless pebble-sized chunks of rubble – rocky and metallic close in, icy beyond the snow line. The planet construction line was underway.

Gradually, through increasing collisions, the great majority of these primordial fragments were deflected towards the mid-plane of the disc. With the fragments thus concentrated into a thinner plane, the rate of collision and agglomeration in the disc then escalated drastically. After only another 1000 years or so, the primordial pebbles had grown to dimensions of several kilometres forming mountain-sized 'planetesimals'. This marked a turning point in planet construction.

Because of their dimensions, the planetesimals now grew not only by collisions with other fragments, but also by virtue of their own gravity. The larger the planetesimals became, the more matter they attracted. And so, only 10 000–100 000 years after the appearance of the Solar Nebula, the inner disc overflowed with innumerable bodies ranging in size up to that of the modern Moon. These bodies, quite justifiably, are known as 'protoplanets'.

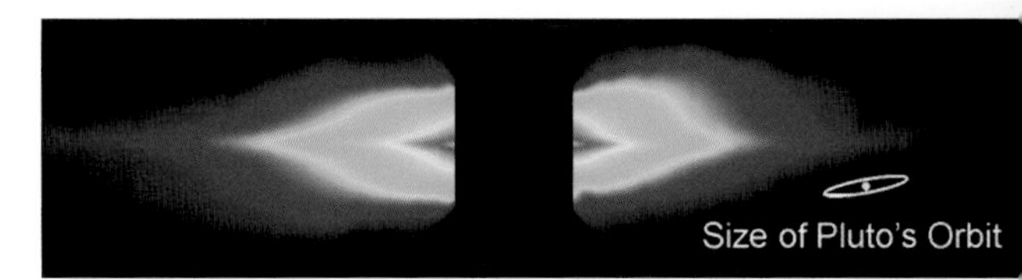

Image above: Some 59 light-years from Earth lies a star known as Beta Pictoris. Since 1984, astronomers have known that Beta Pictoris is surrounded by a pancake of warm gas and dust. Already the material in this disk, which appears to us edge on and is falsely coloured blue and green in this Hubble photo, has started to lump together to form rocky grains and maybe even planetesimals. *Courtesy A. Schultz (SCS/STScI), A. Heap (GSCF/NASA), and NASA.*

2–3 million years
Gas Giants and Asteroids

Not all of the protoplanets grew at the same rate. On the snow line, ices were about ten times more abundant than the silicates and metals closer in. Ices are also very adhesive: calculations have shown that they are 20 times stickier than silicates at comparable impact speeds. Thus, with such a wealth of condensed, gluey materials to work with beyond 5 AU, the agglomeration process operated extremely efficiently there. The end product was the first planet to form: Jupiter.

In less than 100 000 years, a protoplanet larger than the modern Earth appeared on the snow line, a gigantic ball of ice and rock. But its growth didn't stop there, such was the amount of ice. Eventually this icy protoplanet became so large, maybe 15 Earth mases, that it began to suck in even lightweight materials – the gases, principally hydrogen and helium, that still form the greatest part of it today. In this way, the proto-Jupiter gorged itself for several hundred thousand years, after which time it had swept a clear path for itself in the disc. As the planet orbited the Sun, it sucked in gas from either side of the gap it had created, and gradually the reservoir that spawned it began to run dry. What finally stopped Jupiter's growth in its tracks, though, was not a lack of raw material. It was the Sun. After Jupiter had been growing for about one million years, maybe less, the contracting Sun entered the T-Tauri phase. Its powerful wind surged through the Solar Nebula like a tsunami and blasted the unused gas away,

Image below: Far out in the Solar Nebula orbits a giant icy planetesimal some 20 times the mass of the Earth and growing (right). Its enormous gravity draws in gas from either side of a gap in the Solar Nebula as the planet – destined to be Jupiter – clears a path for itself.

deep into interstellar space. At last Jupiter's growth was quenched. But by now it had hoarded more than 300 Earth masses. Unable to grow any larger, the giant planet – by now surrounded by its own gigantic disc of gas and dust, similar to the Solar Nebula itself but on a smaller scale – settled down and began to cool. This was about 3 million years down the planet-production line, long before any of the other planets appeared, with the possible exception of Saturn.

This early appearance of Jupiter spelt trouble for those nearby planetesimals that had not been swept up in the planet's formation. Those that passed close to Jupiter experienced a tug due to the planet's gravity. Over time, some of these planetesimals developed chaotic orbits and were flung out of the disc. Those that remained, unable to group together because of the constant bullying of Jupiter's gravity, survived until the present day in the guise of the asteroids. We shall learn more about these bodies in Part 3.

Saturn, a gas giant similar to Jupiter, came about in a similar manner. But, being twice as far from the Sun, its ice and rock core took longer to form in the relatively sparse surroundings. By the time the solar wind turned on and blasted away the unused gas, Saturn had not had enough time to grow as large as its cousin. A similar fate would meet the next two planets to form, several million years later: Uranus and Neptune.

3–10 million years
Ice Giants and Comets

Image opposite: As a giant protoplanet pulls gas from the Solar Nebula, the stolen material swirls around and forms a disc like the Solar Nebula but on a much smaller scale. In this image the planet Uranus is growing at the centre of its disc, now detached from the larger Solar Nebula. The Sun, still contracting, is at top right.

By about 3 million years, Jupiter and Saturn had formed and were cooling down. But the protoplanetary disc was still very active. Closer to the Sun, the rocky planetesimals were continuing to gather. And much further from the Sun – twice as far out as Saturn is, and beyond – so too were the last of the icy planetesimals. Despite the abundance of ice there, it took longer for icy protoplanets to accrete to the dimensions where, like Jupiter and Saturn, they could pull in gas directly from the disc, because the orbital speeds there were slower. Eventually, though, two more dominant protoplanets of ice and rock did develop. These would become the outermost giants, Uranus and Neptune.

In time these kernels of rock and ice, each about as massive as the modern Earth, began to stockpile hydrogen and helium, just as the larger cores of the gas giants had done a couple of million years earlier. But they had arrived on the scene too late. The Sun was by now past its T-Tauri phase, and very little gas remained in the protoplanetary disc. For a few more million years Uranus and Neptune seized what little gas they could from the ever-diminishing supply, but their growth ceased after about 10 million years – the exact time remains uncertain. The end result was a pair of planets a little over one-third the diameter of Jupiter and only 5 per cent of its mass. And yet, despite their diminutive statures compared with Jupiter, Uranus and Neptune are each still heavier than 15 Earths. They were more than capable of joining in the game of cosmic billiards demonstrated earlier by Jupiter and Saturn. While Uranus and Neptune were still forming, those icy planetesimals that they could not sweep up were instead tossed away like toys that no longer pleased. Today, these fragments, known as comets, surround the Sun in two extensive reservoirs. One, the Kuiper belt, extends a little beyond the orbit of Neptune and is constrained largely to the plane of the Solar System; these fragments are also known as trans-Neptunian objects. Meanwhile, much, much further out, trillions more comets orbit the Sun in a gigantic spherical shell known as the Oort cloud, perhaps more than a light-year in diameter.

In some respects, Uranus and Neptune are like Jupiter and Saturn, but without those planets' gaseous mantles of hydrogen and helium. And so, with a much smaller gas content compared with the proportion of icy substances such as water, methane and ammonia, Uranus and Neptune are not true gas giants. They are best referred to as the ice giants.

3–10 million years
Regular Satellites

Image opposite: As the circumplanetary discs continue to feed material into the planets growing at their centres, the rest of the material in the discs lumps together to form the building blocks of satellite systems. In this depiction the four regular moons of Jupiter are emerging from the disc that surrounds that planet.

While the four giant planets were forming, they were not doing it alone. As each of the giant protoplanets stole gas from the Solar Nebula, the material had swirled around the icy kernels to form gas discs like the Solar Nebula on a much smaller scale. Exactly as in the Solar Nebula itself, the particles in these discs had begun to lump together into larger building blocks – and new, independent worlds had started to appear in orbit around the planets. These would become the giant planets' satellite systems – their moons. Because these moons formed from discs, like the planets, they now tend to orbit their planetary hosts in a thin plane, each in the same direction as the others and in fairly circular paths. Moons with these orbital characteristics also tend to be large. They are known as regular satellites.

It is probable that the regular satellites grew to maturity very quickly, even before their planets did. Why? Simply a question of scale. The discs that surrounded the newly emerging giant planets were much smaller than the Solar Nebula, so they had correspondingly shorter orbital timescales. Their rich cargoes of icy volatiles grew to protoplanet dimensions much more quickly than the planets did. But not all of the moons formed at the same time. The Jovian disc, right on the snow line, would have been the richest. So Jupiter's regular satellites – Io, Europa, Ganymede and Callisto – no doubt formed first, alongside their planet, at T-plus 2–3 million years. These are known today as the Galilean moons, after their discoverer. The next moons to form were the seven or eight largest satellites of Saturn, followed by Uranus' biggest five, and finishing with the moons of distant Neptune several million years after the appearance of the Galileans. Today, however, Neptune does not have a regular satellite system. It is possible, as we shall see later, that its original moons were destroyed when Neptune's gravity netted a rogue protoplanet called Triton. This worldlet went into a retrograde, or backwards, orbit around Neptune and collided with or gravitationally ejected those moons already present. Triton remains today as Neptune's only large satellite, though it is not regular because it did not accrete in a disc around that planet. Triton is a so-called irregular satellite, one of many found in orbit not only around Neptune, but also around all of the other giants. Triton aside, these irregular moons are mostly small lumps of ice and rock that were captured by the planets long after they had formed.

At last, with the giant planets, the regular satellites, the asteroids and the comets in place, the outer regions of the Solar System quietened down. Ten million years had passed. But there was a long way to go. Closer to the Sun, the planet-building factory was still in full swing. There, playing catch-up, the terrestrial planets were emerging.

10–100 million years
Terrestrial Planets

The terrestrial planets were latecomers. Because ices could not condense near the Sun, the materials (rock and metal) from which these planets coalesced were a lot less abundant than those that formed the giants further out. So, while the gas planets had formed within a million years – or at most a few million years – and the ice giants took maybe ten million years, for the terrestrials the formation process was even longer.

At least the initial growth of the terrestrial planets, within a few astronomical units of the Sun, had been very fast. Once the first rocky planetesimals had appeared, they had begun gravitationally to attract smaller bits of nearby debris. As we have seen, these first planetesimals grew to dimensions of hundreds or thousands of kilometres in less than 100 000 years. After about one million years the innermost regions of the Solar Nebula were populated by several large rocky and metallic protoplanets approaching the size of Mercury. And by 10 million years these protoplanets had grouped together through gravitation so that only four dominant spheres remained. These, at last, were the primitive terrestrial planets: from the Sun outwards, Mercury, Venus, Earth and Mars. But even after all four of the giants and their satellites had emerged, the terrestrial planets had grown to only half their eventual masses. And they had a very long way to go to make up that missing half – because the supply of available fragments in the disc was now much lower. Moreover, the terrestrial protoplanets had become large enough for the addition of more planetesimals to have a smaller and smaller effect on their size as they continued to accrete. Thus the growth of the terrestrial planets slowed very significantly.

Tens of millions of years after Neptune and Uranus had formed in the frigid, far reaches of the protoplanetary disc, even after the Sun had started on the main sequence, the terrestrials kept on growing, more and more slowly. In total, it took perhaps 100 million years for the terrestrial planets to mop up the debris, double their masses and swell to their present diameters. But because they never did grow large enough to pull in discs of gas from the Solar Nebula, not one of the terrestrial planets has any regular satellites. (We shall see, however, that Mars did capture two planetesimal moons, and that the Earth's Moon is a special case.) Earth and Venus ended up with roughly equal masses, while Mars acquired only a little more than a tenth of that mass – later we shall see why. Meanwhile, we shall learn that Mercury might have started off with more mass, but lost much of its outermost regions in a gigantic collision with another protoplanet.

Image left: Close to the Sun, the terrestrial planets are emerging. Here, the planet Earth – still molten – is approaching its modern size as it slowly mops up the remaining debris in its vicinity.

100–1300 million years
The Heavy Bombardment

Image opposite: Seen from orbit, the primitive Earth and its recently formed Moon endure the bombardment that, 3800 million years later, is still evident on their surfaces – especially on the Moon. This image shows the process in its early stages, when the bombardment was at its peak.

The planets had finally finished growing. Now they would begin their long process of evolution towards the way we see them today. By now, about 100 million years had passed and the Solar Nebula was relatively sparse. Yet its activity did not stop completely. For the Solar System was still littered with fragments of debris that had not yet been ejected from the system by the giants or been swept up by the terrestrials. It was at this point that the Solar System entered what astronomers call, quite justifiably, the heavy bombardment phase.

For hundreds of millions of years, leftover scraps continued to rain down on the planets and their satellites. This is the battering that shaped the planets' and moons' crusts, and the majority of it occurred in the first 600 million years or so of their creation. A glance at the surface of the Moon gives ample reminder of this violent phase in the Solar System's history. Many of the craters there are well over 100 kilometres across. One of them is about 12 kilometres deep and 2500 kilometres across – greater than half the Moon's diameter. Called the Aitken basin, it is the largest known impact structure in the entire Solar System, carved out when the Moon was struck a glancing blow from a piece of rock and metal some 200 kilometres across. This constant barrage meant that the crusts of the terrestrial planets and moons oscillated between molten and solid states for many hundreds of millions of years. The heaviest elements sank to their centres, while the lighter substances, buoyed up, stayed near the surfaces. In this way the terrestrial planets and the satellites developed differential structures: in the planets, crusts and mantles of rock now surround molten cores of denser metal; and in the moons, the central cores are primarily rocky, with lightweight ices fashioning the mantles and crusts.

The early Solar System saw troubled times. But gradually, as more and more planetesimals collided with the planets and satellites, and were thus removed from the scene, the cratering rate began to drop. Some 1200 million years after the last of the planets had appeared, the craters were occurring perhaps 30 or 40 times less frequently than they had been 400 million years earlier. This point in history, about 3300 million years ago, marked the end of the heavy bombardment phase. The cratering did continue after this, but at a more or less constant although substantially reduced rate.

It was during the last few hundred million years of the heavy bombardment that the planets and satellites of the newly formed Solar System, after aeons of turmoil, began to develop their atmospheres.

700–1300 million years
Building the Atmospheres

Most if not all of the planets developed primitive atmospheres while they were still forming. The giants, as we saw, got their hydrogen–helium atmospheres by pulling in these gases from the Solar Nebula, and these have remained essentially unchanged since. Similarly, the terrestrial planets scooped thin veils of hydrogen and helium from the protoplanetary disc as they moved around within it. But these planets, having much punier gravitational pulls than their giant cousins far from the Sun, were unable to retain these lightweight, primitive skies. Slowly, they slipped away into space, their loss hastened along by the Sun's solar wind.

Gradually, though, as the rate of impacts in the inner Solar System dropped after several hundred million years, the terrestrial planets started to cool. It was during these cooling stages that they developed their secondary atmospheres, via a process known as outgassing. These new skies came from the planets themselves. How? All rocks contain traces of compounds such as water or carbon dioxide that are chemically sealed within the mineral structure of the rock. When these rocks are heated sufficiently, those chemical bonds begin to sever and the trapped gases are released. The terrestrial planets were molten and extremely hot after they had first formed. And so, over hundreds of millions of years during the heavy bom-

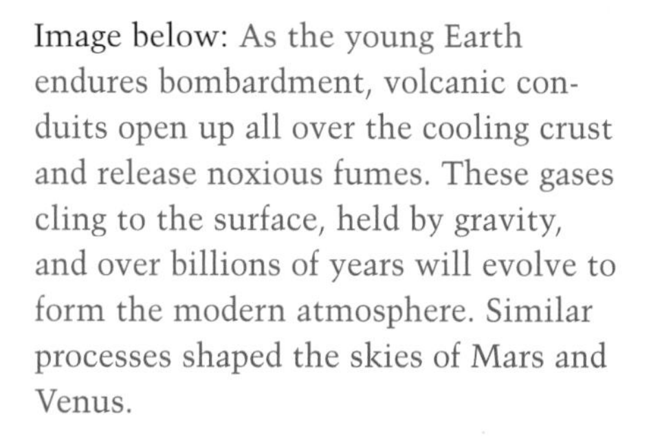

Image below: As the young Earth endures bombardment, volcanic conduits open up all over the cooling crust and release noxious fumes. These gases cling to the surface, held by gravity, and over billions of years will evolve to form the modern atmosphere. Similar processes shaped the skies of Mars and Venus.

bardment phase, these hot balls of rock began to release their locked up vapours through volcanic fissures as they started to cool. Carbon dioxide, carbon monoxide, nitrogen, water vapour, and perhaps hydrogen sulphide were released in this way. In addition to outgassing, planetesimals from the orbit of Jupiter and beyond ventured regularly into the inner Solar System, thrown inwards by the mighty gravities of the giant planets far from the Sun. These comets and asteroids no doubt added a significant water content to the planets' atmospheres – and in fact helped to seed the oceans on Earth.

The secondary atmospheres were in place within several hundred million years of the formation of the planets, while they were still sustaining heavy bombardment. As a result of that bombardment, lightweight Mars ultimately lost 99 per cent of its original secondary atmosphere, which was blasted away into space. And neither Mercury nor the Moon could retain their secondary atmospheres because they did not have sufficient gravity to hold on to even the slow-moving, heavy gases. Over time, all of the planets' atmospheres have evolved. Today, Venus' atmosphere is 100 times more substantial than Earth's, which in turn is 100 times more substantial than that of Mars. But these are stories for Part 3.

4500 million years? Formation of the Ring Systems

Image opposite: A comet has encroached within Saturn's gravitational danger zone and been torn into fragments. With successive passes of Saturn some of the fragments impact the planet, while others go into orbit. There, ground down into rubble as they collide among themselves, they have already started to form a ring system that might last hundreds of millions of years.

With the emergence and subsequent evolution of the planetary atmospheres, the Solar System was almost complete. Only two things remained to be added: the rings of the giant planets, and some of the smaller, irregular satellites. The irregular satellites were probably acquired early in the history of the Solar System, when the giant planets captured icy planetesimals from the thinning Solar Nebula. Some are no doubt of more recent origin. The origins of the rings, however, are more difficult to pin down.

The most famous ring system is Saturn's. Consisting of countless boulder-sized, and smaller, icy chunks in individual orbits about the planet, the rings are exceedingly thin – with relative dimensions like those of a sheet of paper the size of a football pitch. But Saturn is not alone, because each of the other giant planets has similar accoutrements, albeit with different characteristics. Indeed, research has shown that no two systems are alike: they differ from each other in terms of diameter, brightness, and in the sizes and compositions of the particles that constitute them. This is a clue to their formation. But the biggest hint is that most of the rings surround their planetary hosts inside their respective 'Roche limits'. This is the distance from a given planet at which gravitational forces tear apart any body held together mostly by gravity. These clues could mean that the rings are the unassembled ruins of moons that strayed within this danger zone and got ripped to shreds, or the remains of comets that got too close and suffered a similar fate. Such a hypothesis neatly explains the differences in the rings: they depend on the constituents of the bodies that were destroyed in their making. Alternatively, the rings could be relics from the discs that surrounded the giant planets in the early Solar System, from which the regular moons formed. But this is unlikely. First, Saturn's icy particles would have evaporated long ago while the planet was still a hot ball of gas. Second, computer simulations of particle orbits suggests that ring systems are unstable over long periods of time.

If these dynamical studies are correct, then the rings are of relatively recent origin – probably dating to less than 100 million years ago. But even this has its problems. How is it that we are alive at just the right time to witness the existence of not just one, but four ring systems, if they are all transient? The best answer is that the rings have existed for longer, but their particles are continually replaced by the break-ups of small moons and comets.

4660 million years
The Modern Solar System

Image below: The Solar System as it appears today. In this representation, the sizes of the planets and their distances from the Sun have been exaggerated for clarity and are not to scale. Only the outermost planets can be seen clearly – the terrestrials are huddled much closer to the Sun and are lost in its glare. If the Oort cloud were shown on the same scale as the outer planet orbits here, it would span the best part of a kilometre.

At last we come to the present day. We have journeyed over 4 billion years in time to get to where we are now. But as we peer out into the depths of the Solar System that is our home, we can easily see the evidence of its formation. We see near-circular orbits, most of which lie in the same plane – a relic of the Solar Nebula. We see worlds with battered surfaces – the scars that betray the long and troubled period of meteoritic rain known as the heavy bombardment. And, because of the way the Solar System was made, we can now count five distinct zones within it.

The first zone lies within 1.7 AU of the Sun. This is home to the four terrestrial planets, Mercury, Venus, Earth and Mars. These are small worlds of rock and iron, forged from the hottest fires of the Solar Nebula. The expanse from about 2–3.3 AU marks the second zone, that of the asteroids. Some of these stony or metallic bodies have not been modified extensively in over 4 billion years, which means that they contain some of the most primitive materials in the Solar System. Zone three is much larger, the realm of the giants. Its innermost boundary is marked by the planet Jupiter, almost twice as far from the Sun as Mars is; its outermost

boundary lies at Neptune, fully six times further from the Sun than even Jupiter. All of the giant planets are far bigger than the terrestrials, with compositions of ices and gases – and comparatively little rock. The fourth zone is the Kuiper belt of comets, or the so-called trans-Neptunian objects. This extends from roughly the orbit of Neptune to an unknown distance, but perhaps as far as 1000 AU. This icy wasteland is also home to the tiny worldlet known as Pluto, which we will meet in Part 3. The fifth and last zone in the Solar System is the largest by three orders of magnitude. It is the spherical shell of icy comets called the Oort cloud that surrounds the Sun at a distance that might even exceed 50 000 AU – a large fraction of a light-year. The comets in both the Oort cloud and the Kuiper belt owe their presence to the gravities of Neptune and Uranus.

And so we come to the known boundary of the Sun's family. Somehow it is fitting that the phantom Oort cloud now surrounds the Sun on such a vast scale. It is similar in scale to that of the frigid globule of gas and dust from which everything in the Solar System sprang so long ago.

Part 3
Solar System Past and Present

'Everything flows and nothing stays.'

Heraclitus, *Cratylus* (Plato)

It is now more than four and a half billion years since the Solar System came into being. Generally there has been little evolution in the grand scheme of things. Along with some of the planets' moons, the asteroids have surfaces that have not been modified extensively since the heavy bombardment stopped, 3300 million years ago. They still populate the belt between Mars and Jupiter. The comets still surround the Sun in the Oort cloud, and the planets' orbits are essentially unchanged.

But on local scales it is a very different story. Since the initial fires of their births so long ago, the planets, their moons and even the Sun have seen significant changes. We will look at these in this, the third part of the book. Starting with the Sun and working our way outwards, we will investigate each of the elements of the Solar System in turn to see what they are like now – and how they have come to be that way. We shall see that the Sun is slightly larger and a bit more luminous today than it was when it was born. It will become evident why the Earth became the only place capable of supporting life, and why it no longer shows signs of its formation. We will discover a different Mars, a waterworld like Earth, and learn how it evolved into the cold and barren desert it is today. And there are other sights too: a larger Mercury than exists today; volcanoes that have changed planets' surfaces beyond recognition; and moons that have been shattered in cosmic impacts and later reformed into new and unusual forms. Having seen where the Solar System came from, it is time now to see how it has evolved.

The Sun – Local Star

The Sun is an ordinary yellow dwarf star. It is a gaseous, radiating globe 1 392 000 kilometres across, easily big enough to contain the entire orbit of Earth's Moon. Like all other stars, the Sun stays balanced against gravity, in a state called hydrostatic equilibrium, because it has a nuclear reactor in its heart. There, hydrogen nuclei are smashed together to make helium, and radiation is released, at a high enough pressure to hold the star up against gravity. This stage of the Sun's existence is known as the main sequence. In the 4600 million years since it was formed, the Sun has changed very little in terms of general structure. But there are some subtle differences in rotation rate, luminosity and diameter that are a consequence of its solar wind, and the way in which it produces energy.

Sun Data

Mass: 1.99×10^{30} kg or 332 946 times Earth's

Diameter: 1 392 000 km or 109.1 times Earth's

Surface gravity: 27.9 gees

Surface temperature: 5700 Celsius

Core temperature: 15 000 000 Celsius

Mean rotation period: 25.4 days

Spectrum: G2 V

Distance from galactic centre: 25 000 light-years

Orbital speed in galaxy: 220 km/s

Orbital period in galaxy: 225 000 000 years

Anatomy of the Sun

Image opposite: Seen from a distance of 210,000 kilometres outside Mercury's orbit, the Sun and its innermost planet share the same apparent angular dimension. This image shows an imaginary view of Mercury from exactly this distance, eclipsing the Sun directly behind it just as the Moon does on Earth. Such an alignment affords a spectacular view of the Sun's outermost atmosphere, its corona, otherwise invisible due to the Sun's glare.

Because the Sun is composed entirely of ionised gas – in which intense radiation has stripped the atoms of their electrons – its mean density is very low. On average our star is only 1.4 times denser than water. Most of this material, more properly known as a plasma, is hydrogen, 71 per cent by mass. Helium accounts for 26 per cent, and the rest is chiefly oxygen, carbon, nitrogen, neon and iron, also in gaseous form. Despite the low average density, though, the Sun's local density varies enormously. Not surprisingly, the densest part is the centre, crushed under the weight of all the layers on top. Here, the density is more than 13 times that of solid lead, and the temperature is 15 million Celsius. This is the Sun's engine, where hydrogen is processed into helium. It is known as the core. It accounts for about the innermost 20 per cent of the Sun but contains well over half the star's entire mass because the material is so highly compressed. And yet despite this density the core is still gaseous. Ions roam as freely there as atoms do in a gas.

Just outside the core is the radiative zone. This extends out to 70 per cent of the Sun's radius. As the name suggests, the radiative zone is the region where energy is transported away from the core in the form of radiation. But it takes a very long time. Even here the Sun is so dense that the little packets of radiation, called photons, travel only a few millimetres before they hit an ion and bounce off in another, totally random direction. Because of this it can take individual photons 170 000 years to travel the length of the radiative zone.

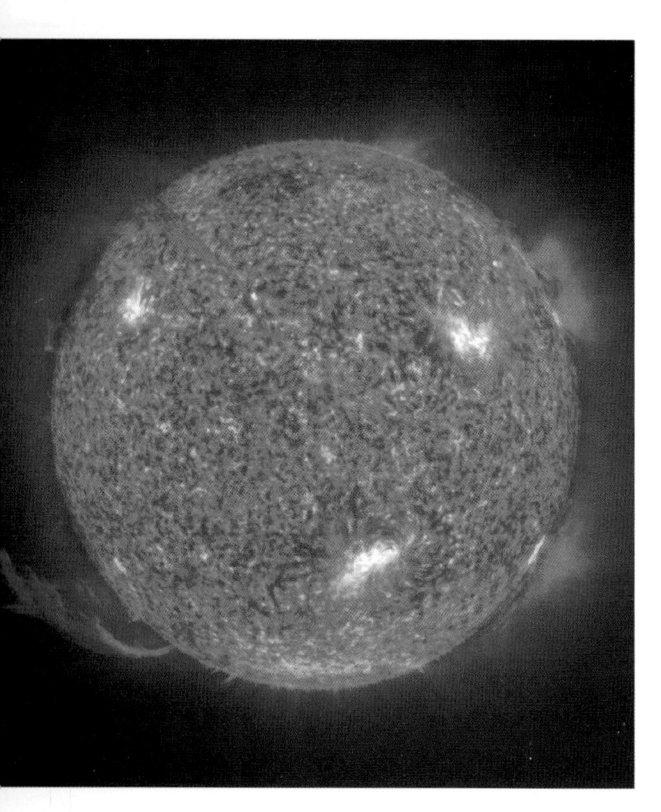

Image above: This image, taken in false colour by the Extreme Ultraviolet Image Telescope on board NASA's SOHO observatory, shows a vast eruptive prominence on the bottom left limb of the Sun. The gas temperature in the prominence is around 70 000 Celsius. *Courtesy SOHO/EIT Consortium.*

Image opposite: The Sun has three distinctly different zones inside it. The energy-generating core occupies the innermost 20 per cent or so (white). Just above that is the radiative zone, which extends 70 per cent of the way to the surface (yellow). The bubbling convective zone fills the outermost 29 per cent (orange). And the visible 'surface', called the photosphere, is only 500 kilometres thick (cream).

Eventually, though, the photons do reach the edge of the radiative zone – only to encounter another barrier. This far out in the Sun, the temperature is only around 2 million Celsius, and the gas is quite opaque. Thus, as the radiation streams out of the top of the radiative zone, it grinds to a halt, absorbed by the ions in the cooler gas there. The energy must find another way to escape to the surface, and it does this with the aid of convection. As more and more photons arrive at the outer edge of the radiative zone, the gas there becomes slightly hotter than the material immediately above. Warm gas is lighter than cold gas. And so the heated gas gets buoyed up just like a layer of warm water lying on top of colder water in a bath. Suddenly the heated plasma is displaced upwards in huge units, hundreds of thousands of kilometres across. These packets of gas rise steadily until, after about a week, they near the outermost edge of the Sun. Because of the way in which the Sun's energy is transported to the surface in this its outermost third, this region is known as the convective zone.

The rising cells of gas, pushed up through the convective layer, produce the eventual image of the Sun that we see. As the cells arrive, they spread out with the pressure of more gas coming up underneath them, and their cargoes of gas cool down and sink, later to be reheated. The result is the 'pebble-dash' appearance that the Sun has in high-resolution photographs – the so-called solar granulation. This, the visible surface of the Sun, is only about 500 kilometres thick – thinner than tissue wrapped around a bowling ball. It is called the photosphere, a name that means 'sphere of light'. For it is here that the Sun's material becomes transparent to optical radiation for the first time. At last, the photons emitted here can travel unimpeded. They stream out into space where, 8.3 minutes later, we detect them on Earth as sunlight. But it is wrong to think of the Sun as having a solid surface. The gases in the photosphere have only 10 per cent of the density of the air we breathe. They merely seem solid because we cannot see beneath them – the gases there are too absorptive.

The Violent Sun

Aside from granulation, the Sun's surface has other features. Most notable are its spots. Sunspots are a result of the Sun's magnetism. This, in turn, is a result of rotation and convection. Because the Sun is not a solid body it rotates differentially: once in 26 days at the equator and once in 30 days at the poles. As a consequence, the Sun's magnetic field lines, somewhat like those of a bar magnet, get distorted and stretched as the star spins. Where the distortion is greatest, the magnetic field becomes amplified and this inhibits the flow of local gases. This makes these local regions on the Sun cooler than the rest, and we see them as relatively dark sunspots. On average, sunspots – which can last just hours or months – have temperatures of around 4200 Celsius, which is about 1500 Celsius cooler than the average surface temperature of 5700 Celsius. The largest can be bigger than the Earth.

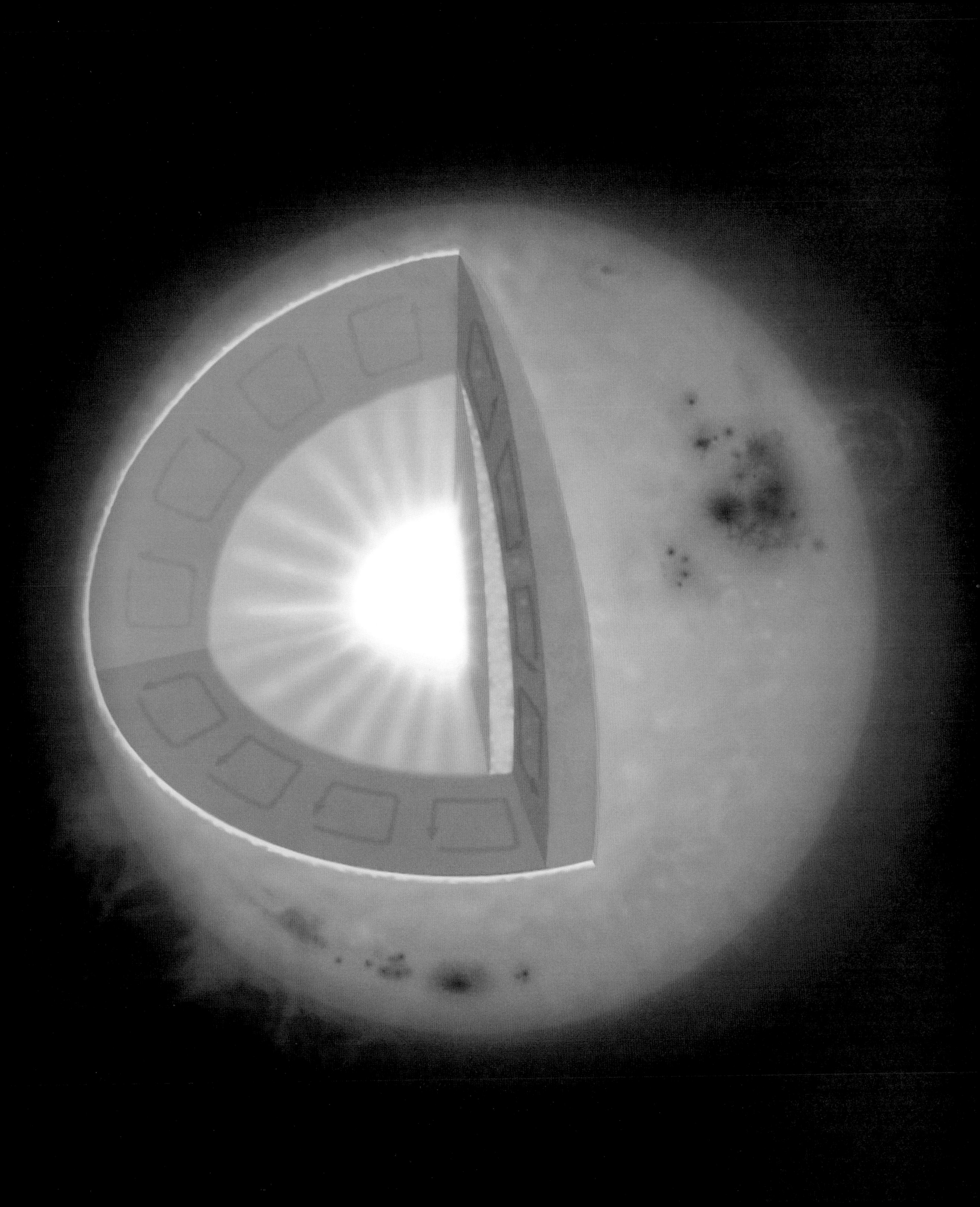

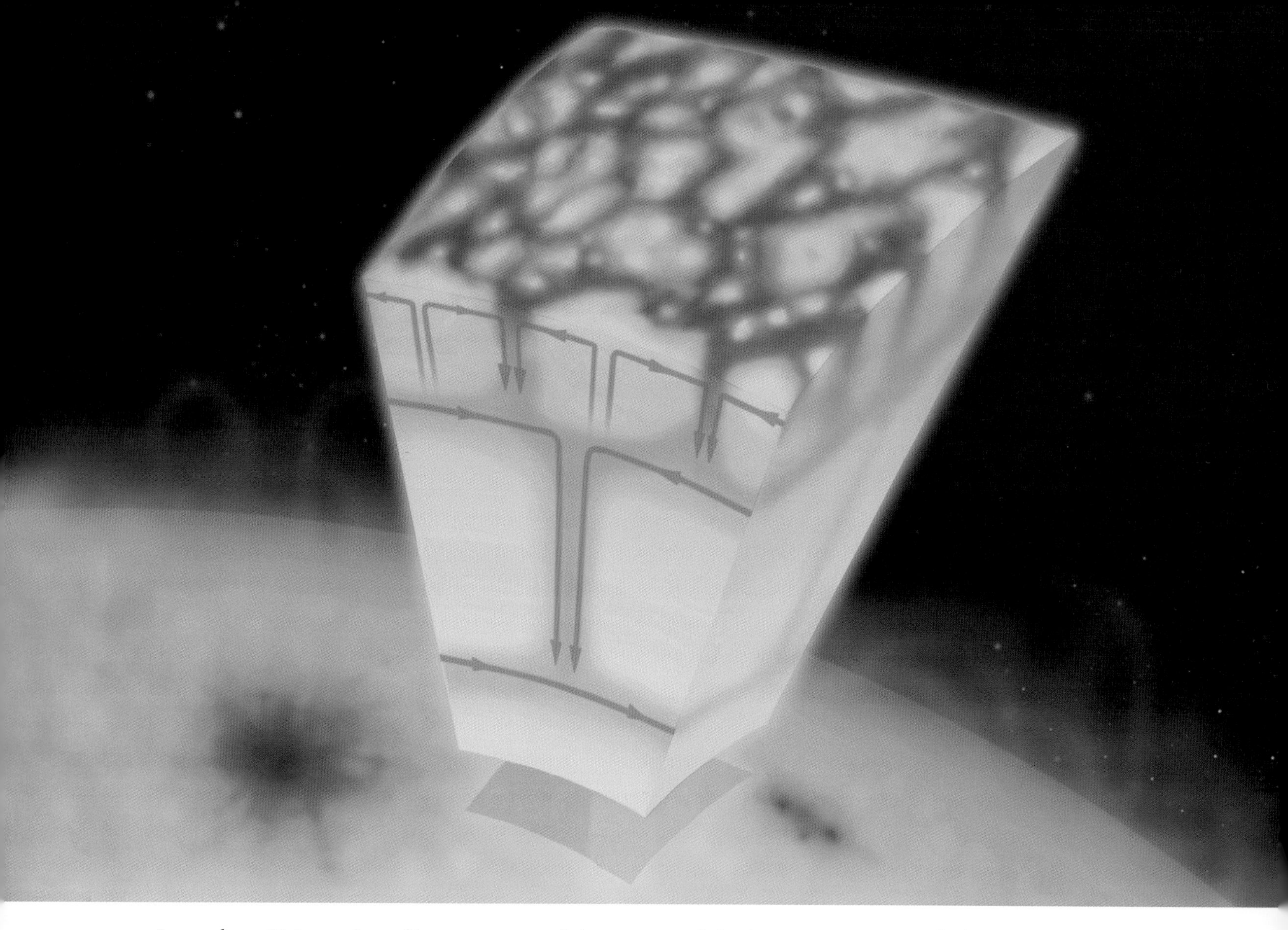

Image above: Rising packets of hot gas in the outermost third of the Sun give rise to the patchwork pattern of grain-like cells on the surface, known as the solar granulation.

Other aspects of the Sun's magnetism include solar flares and prominences. Prominences are great, arch-shaped clouds of gas that hang in the Sun's atmosphere – its 'chromosphere' – tens or hundreds of thousands of miles above the photosphere. They are bunches of gas, denser than the gas around them, entrained in magnetic loops, and have lifetimes of several hours. Flares are sudden explosions that originate near active regions, heating up the gas in the chromosphere and corona – the outermost atmosphere – to temperatures of hundreds of millions of degrees. A typical flare will blast a cloud of charged particles – electrons, protons and other atomic nuclei – away from the Sun at speeds of up to 70 per cent that of light. Also related to the corona are the so-called coronal mass ejections. Nobody understands exactly what drives them, but coronal mass ejections involve the sudden expulsion of up to 10 trillion kilograms of hot plasma at speeds of up to 1000 kilometres per second. They create shockwaves in the Sun's solar wind – its steady stream of mainly electrons and protons – that can seriously disrupt communications on Earth.

The Changing Sun

So the Sun is a violent place. But it is not as active as it used to be. During the millions of years it took to form, the Sun got spun up by the gases that fell onto it from the Solar Nebula. Since the nebula dispersed, however, the rotation has slowed down considerably. This is because of the interplay between the solar wind and the Sun's magnetic field. As the wind blusters into space, some of it gets channelled along magnetic field lines. This redistribution of some of the Sun's mass radially away from it acts to slow it down, just as a ballet dancer slows down when he extends his arms away from his body after a spin. Both the dancer and the Sun are said to be conserving angular momentum – and, in the Sun, the process is known as magnetic braking. The Sun now spins once in 26 days at the equator, compared with once in 8 days or fewer when it was a T-Tauri star. This produces a weaker magnetic field, and is why the Sun is now a lot less magnetically active than it used to be, fortunately for us.

Image above: This negative image shows a coronal loop on the western limb of the Sun, as captured by the TRACE (Transition Region and Coronal Explorer) spacecraft. Coronal loops are regions where high-temperature gas is buoyed up and held in place above the photosphere by magnetic forces. *Courtesy TRACE/Stanford-Lockheed Institute for Space Research.*

There is another way in which the Sun has altered. Because the core has been converting hydrogen into helium for so long, its composition has changed somewhat since the star started to shine. Now, 37 per cent of the core hydrogen has been consumed, replaced with helium, and this has had consequences for the outward appearance of the Sun. Four hydrogen nuclei go into the making of every single helium nucleus. Each helium nucleus exerts the same pressure as a hydrogen nucleus. But because the total number of nuclei in the core becomes fewer and fewer as more hydrogen nuclei are consumed, the pressure there drops steadily all the while the Sun is on the main sequence. As a result, its core is always contracting, growing a bit hotter. This produces more energy that in turn puffs the Sun up a little and makes it more luminous. Over time, the change in luminosity can become quite significant. Already the Sun is about 10 per cent larger than it was when it emerged from its birth cloud, 4600 million years ago. It is also slightly warmer, and about 30–40 per cent more luminous. This trend will stay with the Sun until the end of the main sequence, more than 6 billion years from now.

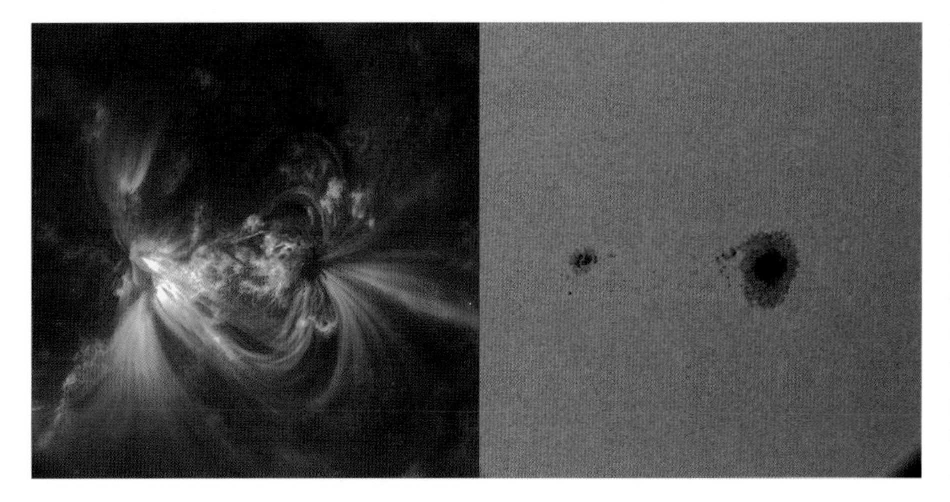

Image left: The Sun can alter its appearance very dramatically depending on the temperature-sensitivity of the imaging equipment. The view on the left reveals million-degree plasma in the Sun's corona, and clearly shows the magnetic field lines heavily laden with high-temperature gas. But on the right, which shows the same scene in ordinary white light, only the photosphere is visible; the sunspots are evidently the focus of the field lines seen in the first image. *Courtesy TRACE/Stanford-Lockheed Institute for Space Research.*

Mercury – Iron Planet

The first planet from the Sun is Mercury, only 40 per cent larger than our Moon. It is a heavily cratered ball of rock and iron, without a satellite of its own, huddled in to the searing Sun more than two-thirds as close as the Earth is. On average, Mercury orbits the Sun at 0.39 AU, once in just 87.97 days. It's a proximity that makes for an exceedingly high surface temperature. Daytime temperatures peak at 430 Celsius at the subsolar point, then drop by more than 600 Celsius at night. No other planet exhibits a more extreme range of temperature. And compared with the orbits of all other planets except Pluto, which are nearly circular, none is more eccentric than Mercury's. The planet's significant iron content also gives it a very high density, comparable to the Earth's. And it is because of its odd orbit and high density that astronomers suspect Mercury's present appearance might be an accident – the outcome of a cosmic collision deep in the planet's past.

Mercury Data

Mass: 3.30×10^{23} kg or 0.055 of Earth's

Diameter: 4878 km or 0.38 of Earth's

Surface gravity: 0.38 gees

Axial tilt: 0.1°

Mean surface temperature: 427 Celsius

Rotation period: 58.65 days

Orbital period: 87.98 days or 0.24 years

Inclination of orbit to ecliptic: 7.0°

Orbital eccentricity: 0.206

Distance from the Sun: 0.31–0.47 AU

Sunlight strength: 4.5–10.4 times Earth's

Satellites: 0

Physical Overview

Image opposite: Mercury sports a battered surface not unlike that of Earth's Moon. Temperatures on the heated face can be hot enough to melt tin. And yet, oddly enough, some of those impact craters may contain traces of ice, brought by falling comets. Some crater floors never see the Sun, and so ice can persist there almost indefinitely in the permanent cold.

Even a cursory glance at Mercury reveals a battered world, not unlike the Moon. The colour is slightly different, though: more coppery than the grey that characterises our satellite. There are no volcanoes, and Mercury is devoid of an atmosphere. It does occasionally capture some gas – helium and hydrogen – from the solar wind, but the density is far too low for it to be considered 'air' in any real sense. Also, liquid water has never existed on Mercury – nor any water at all aside from that brought in small amounts by cometary impacts. This is because the planet formed so close to the Sun that such volatile compounds could not condense. And so, lacking volcanism, air and water – which on Earth are powerful forces of erosion – Mercury's surface is a fossil, geologically dead. It has not changed significantly in several billion years.

For this reason Mercury's main surface features are its impact craters. These are the scars that betray the heavy bombardment the planet endured after it had formed. Craters appear over almost the entire surface and at a wide variety of sizes, forming a mountainous terrain known as the highlands. However, the surface relief in the highlands is not as high as might be expected. Many of the craters are quite shallow. It is as if they have been flooded by ancient lava flows early in the planet's history. Some older craters may even have been buried altogether. Because of this the rolling

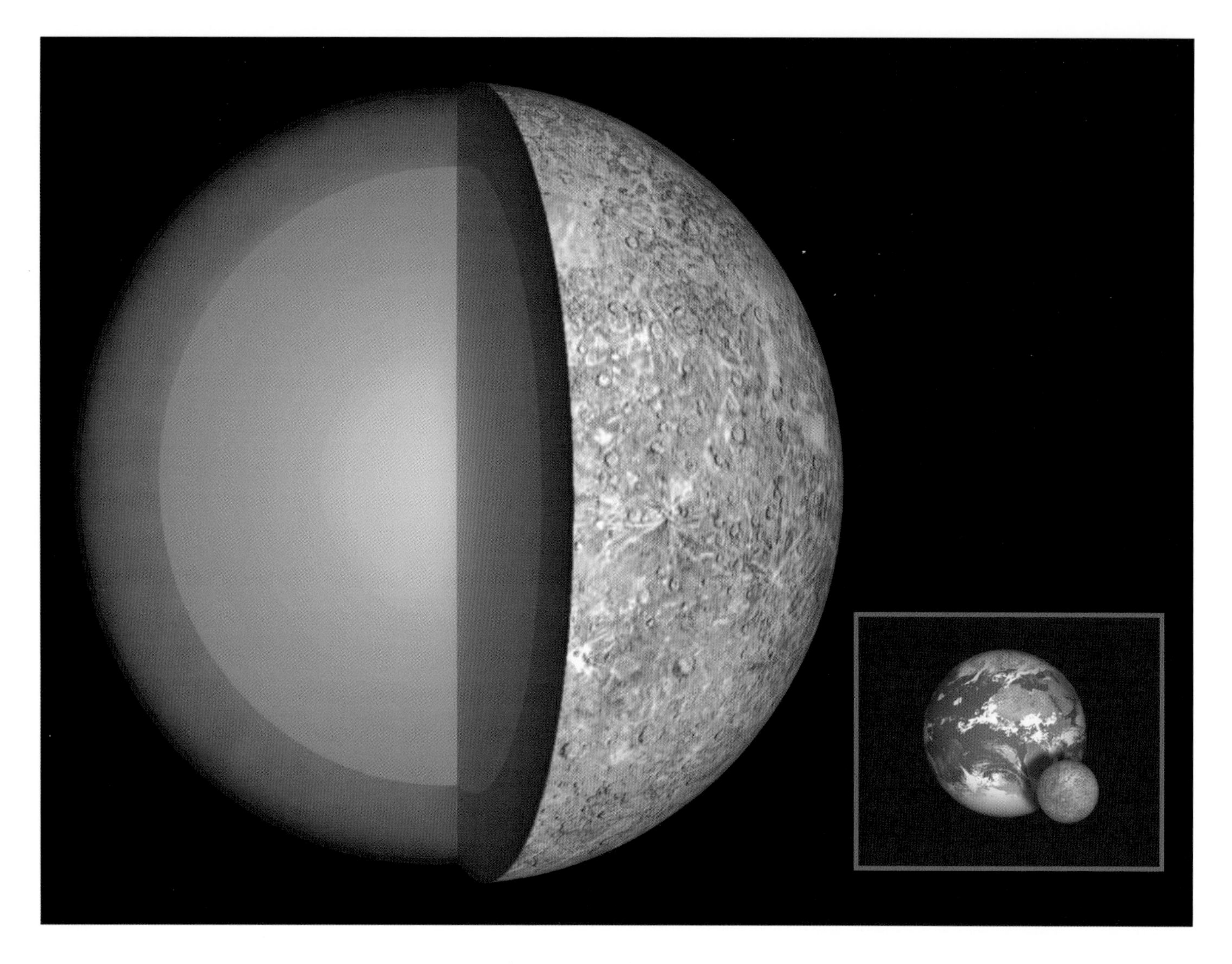

Image above: A cutaway diagram of Mercury reveals an iron core that occupies almost the entire volume of the planet. Part of the core is molten (orange). Above it is a solid silicate mantle about 550 kilometres thick, topped with a solid crust. The inset shows Mercury on the same scale as the Earth.

spaces between individual craters, called the intercrater plains, tend to be somewhat smooth. Meanwhile, there are other, much more extensive smooth areas called lowlands or smooth plains. These are found near the planet's north pole or around and within giant impact features called basins. The largest of these, the Caloris basin, is a vast circular patch about 1300 kilometres across – big enough to contain the British Isles. It was formed when a large planetesimal, 100–150 kilometres in diameter, ploughed into the young planet with the force of a trillion megaton-nuclear bombs. The blast melted the surface locally, forming as it solidified the smooth, round impact scar that we see today.

Aside from craters and basins, Mercury boasts a number of fractures called scarps. Some of these are up to 500 kilometres long and form cliffs that in places jut nearly 4 kilometres into the black sky. They often cut right across craters and basins and are thus more recently formed. Astronomers suspect that the scarps are faults caused by horizontal compression in the crust – evidence, they say, that the planet has gradually contracted and cracked upon cooling. Very likely, therefore, much of Mercury's interior has frozen solid; the body is too small to retain heat for as long as the Earth. But, at the same time, the planet has a measurable

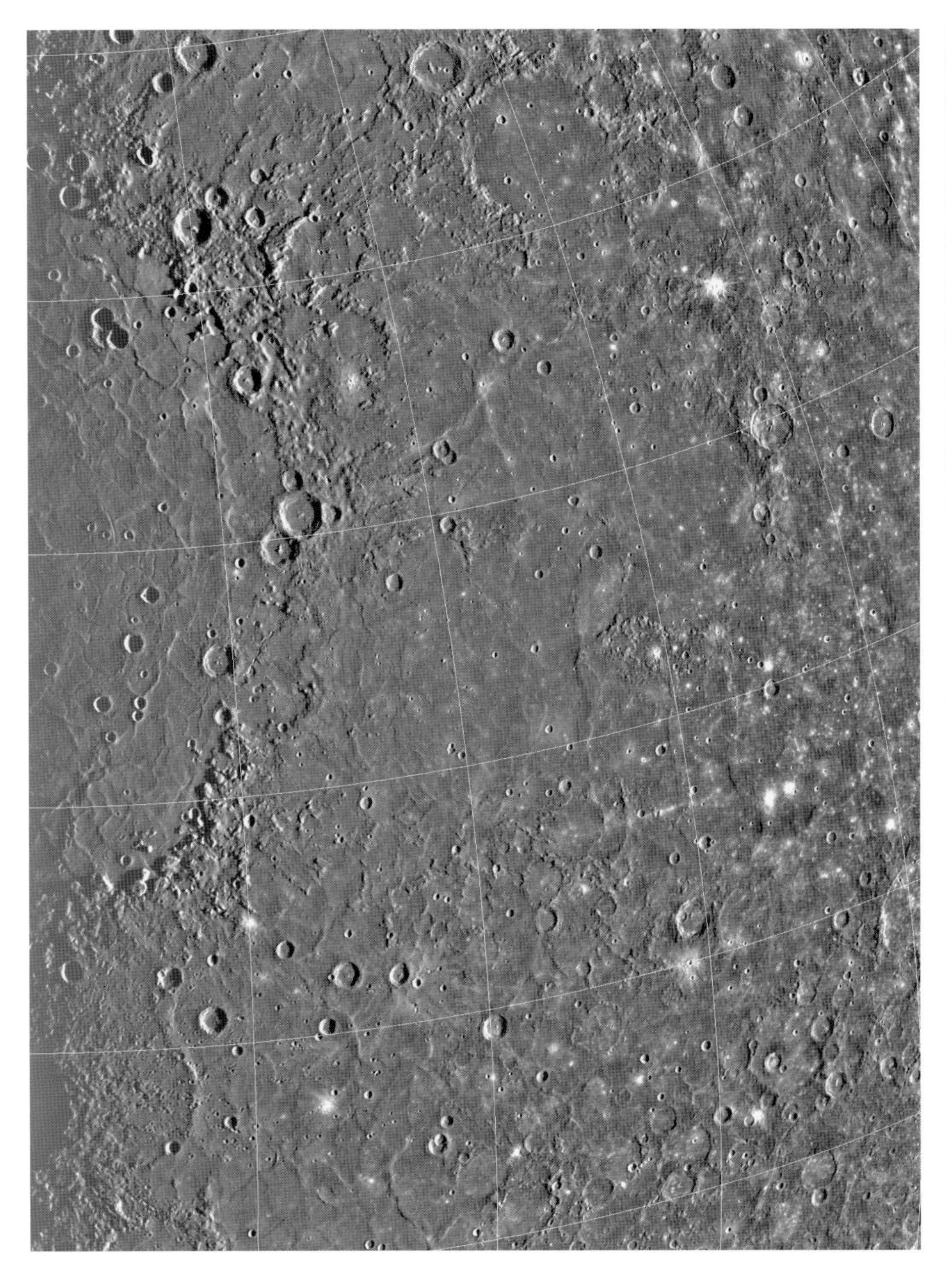

Image above: Just one of many fractures on Mercury, Discovery scarp cuts right across a crater in this Mariner 10 photograph. Scarps are cracks created when Mercury cooled and its crust contracted. *Courtesy of NASA/JPL/Caltech.*

Image left: This Mariner 10 image shows a portion of the Caloris basin, a Mercurian impact structure some 1300 kilometres across. It is one of the planet's most recent features. *Courtesy of NASA/JPL/Caltech.*

magnetic field, and this implies that the interior must at least be partially molten, perhaps in the very centre. Without a fluid interior the planet would lack the convection needed to generate its magnetic field.

Cosmic Casualty

The interior of Mercury must also be very rich in metals such as iron, for the planet has the second highest average density in the Solar System, after Earth. Of course, because it accreted so close to the Sun where only the densest substances could condense, its high metal content is to be expected. But the planet has such a high mass for its size that its iron core must be phenomenal. It extends out to 75 per cent of the planet's radius. This has led some astronomers to the conclusion that Mercury suffered

Image above: The impact that formed Caloris was so enormous that the shockwaves from the event travelled right through the planet and jumbled up the terrain in the hemisphere directly opposite the impact site. This image shows the resultant landscape. *Courtesy of NASA/National Space Science Data Center.*

a cataclysmic collision with another large planetoid while it was still accreting. It is quite conceivable that a large enough impact – with a body perhaps half the size of the young Mercury itself – could have melted the planet's original rocky mantle, jettisoning it into space where it later fell into the nearby Sun. Only the iron-rich core would have escaped annihilation. If this happened late enough, Mercury would not have been able to recover its original mass owing to the rapidly diminishing number of planetesimals in its vicinity. Instead, its growth was stunted, and the result is the planet we see today: dwarfish, dense, and with a relatively odd orbit.

At closest approach to the Sun Mercury is only 0.306 AU from it. But the other end of its orbit lies at 0.467 AU, more than 1.5 times further out. This means that from the surface of Mercury the Sun varies its diameter from two times to more than three times that seen from Earth. The orbit is also tilted with respect to the so-called ecliptic, the centre plane of the Solar System, by 7 degrees. These characteristics may have been imparted to the planet in the very same impact that vaporised its original surface and blasted it into space.

Evolution of Mercury

From the general physical characteristics of Mercury, researchers now think they have some idea how the planet has evolved. Its oldest terrain, accounting for 70 per cent of the surface, is the highlands. The craters there were formed during the early heavy bombardment phase and are thus some 4200 million years old. They date back to a mere 400 million years after the end of accretion in the Solar Nebula. The highland intercrater plains, where some craters have been buried partially or totally, are obviously somewhat younger. These plains are gigantic lava flows that oozed out of the planet's crust on a global scale about 4 billion years ago. Meanwhile, the last major episode of activity on the planet was that which followed the Caloris basin impact. This devastating blow brought more lava to the surface and formed the localised smooth plains or lowlands. These areas have few impact craters and so must have been laid down after most of the heavy bombardment had finished. Thus the Caloris basin is about 3800 million years old – the youngest terrain on the planet. Aside from the cracks brought about as the planet gradually cooled and contracted, Mercury's airless surface hasn't really changed since the Caloris event. The rugged landscape we see now dates to just 800 million years after the formation of the planet itself. It has been frozen solid ever since, for the last 80 per cent of the planet's existence.

There is one other way in which Mercury has changed, though, and that has to do with its rotation period. Currently the planet takes 58.65 days to spin once on its axis. But when the planet first emerged from the busy rubble of the protoplanetary disc it would almost certainly have been going much faster. The cause of Mercury's gradual spin down was the tidal force exerted by the nearby Sun. Because of the way in which gravity diminishes with distance, the Sun's pull on Mercury's surface is much greater than its influence at the planet's centre. Thus in a sense the Sun tries to hold on to Mercury's surface layers as the planet rotates. Yet at the same time, Mercury naturally tries to take its surface layers with it as it spins. This tug-of-war – known as tidal friction – caused Mercury to gradually slow down. As a result it now has the second-slowest rotation period in the entire Solar System. Only Venus spins more slowly than Mercury. And that is the next planet we encounter as we move outward through the Solar System, away from the Sun.

Image above: Ancient, cratered highlands and slightly more recent smooth plains juxtapose in this Mariner 10 close up of Mercury's surface. *Courtesy of NASA/JPL/Caltech.*

Venus – Hell Planet

The Solar System's second planet is Venus, which moves in an almost circular orbit at an average distance of 0.72 AU from the Sun. With a size, mass and density comparable to those of the Earth, Venus has often been thought of as our planet's twin. But in virtually every other way the two planets are about as similar as Heaven and Hell – with Venus a good substitute for Hell. Like Mercury, Venus has no moon. But it does have a substantial atmosphere – a choking shroud of almost pure carbon dioxide. So dense is this waterless sky that it pushes down on the surface at a pressure 92 times that on Earth – equivalent to that at the bottom of a lake 900 metres deep. A global greenhouse effect keeps the surface temperature twice as hot as a domestic oven. And, high above the surface, thick clouds of battery acid permanently block Venus' terrain from optical telescopes. Only radar can penetrate the clouds. The results have revealed a geologically young surface, less than 500 million years old, replete with volcanic features but with relatively few impact craters, compared with the Moon, Mars and Mercury.

Venus Data

Mass: 4.87×10^{24} kg or 0.82 of Earth's

Diameter: 12 104 km or 0.95 of Earth's

Surface gravity: 0.90 gees

Axial tilt: 177.4°

Mean surface temperature: 480 Celsius

Rotation period: 243.02 days or 0.66 years

Orbital period: 224.70 days or 0.62 years

Inclination of orbit to ecliptic: 3.4°

Orbital eccentricity: 0.007

Distance from the Sun: 0.72–0.73 AU

Sunlight strength: 1.9 times Earth's

Satellites: 0

Atmosphere

Image opposite: Even up close, Venus betrays little. Its clouds are so opaque that the planet resembles little more than a featureless, white sphere.

Only slightly smaller than the Earth, Venus is easily massive enough to exert a gravity capable of clinging to an atmosphere. But thick clouds cover 100 per cent of Venus' surface. Early astronomers, unable to penetrate the planet's permanent blanket, had only their imaginations with which to paint this world. They envisaged a lush, fertile swamp-planet, full of strange creatures perhaps resembling alien dinosaurs. They were completely wrong.

Probes have revealed that the chief constituent of Venus' atmosphere is the poisonous gas carbon dioxide, 96.5 per cent by volume. The rest is mainly nitrogen, and there are minute traces of sulphur dioxide, argon, hydrogen and water vapour. This is a substantially different composition from the nitrogen–oxygen-rich skies of Earth. Perhaps the most impressive trait of Venus' dry atmosphere, though, is its incredible density. It harbours enough gas to make almost 100 atmospheres as thin as Earth's. And the weight of that gas is formidable. At the surface, the atmosphere exerts a pressure of more than 90 kilograms on every square centimetre. You'd need a spaceship built to withstand submarine environments just to survive there.

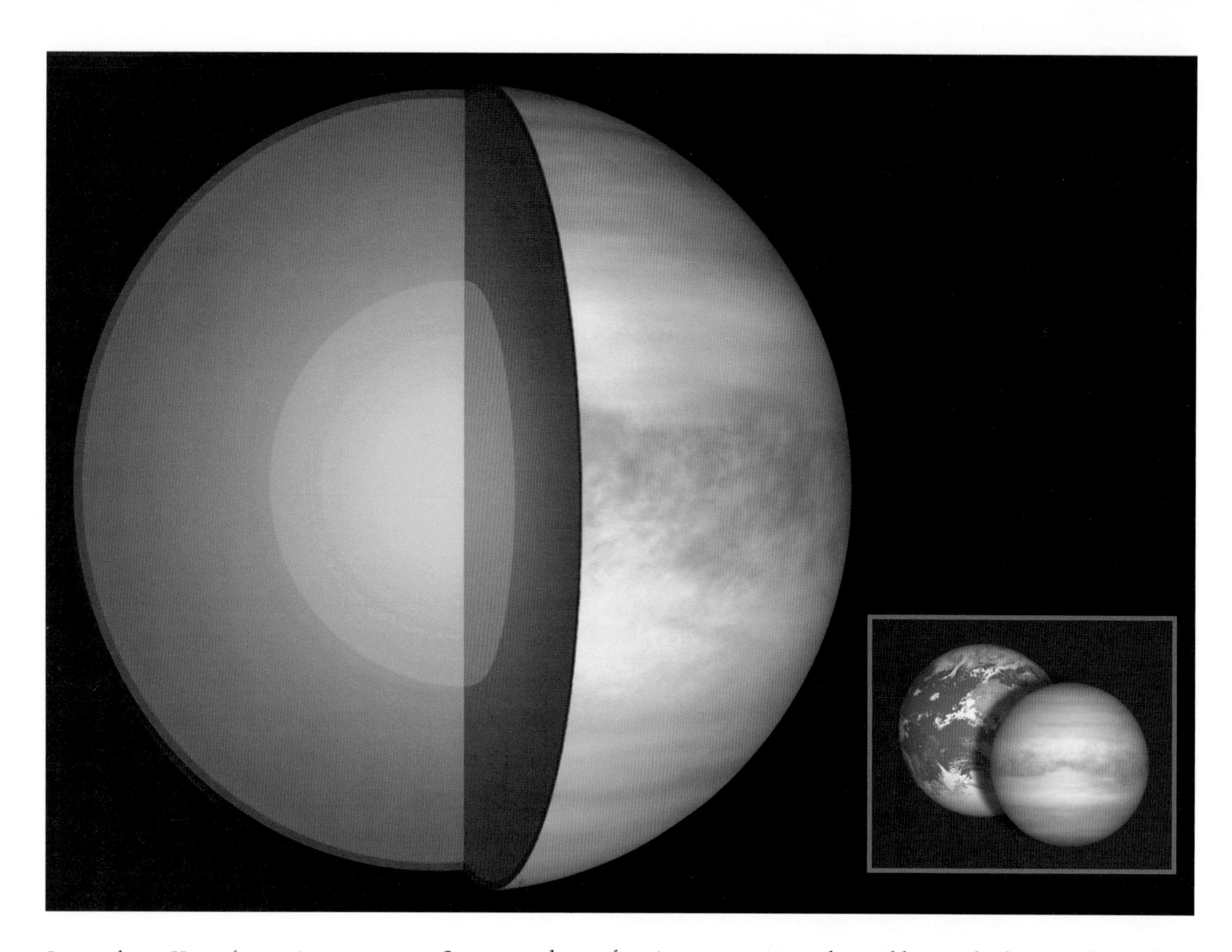

Image above: Venus has an iron core (orange), probably solid, taking up about 50 per cent of its radius. This is surrounded by a thick silicate mantle and topped with a relatively thin crust, 30 kilometres deep. The inset shows Venus on the same scale as the Earth.

Step onto the surface in a spacesuit and you'd be crushed instantly. And, as if that were not enough, the ship would have to be exceedingly well insulated too, to shield its occupants from the devastating temperatures outside. Despite the fact that Venus is much further from the Sun than Mercury, Venus has the hottest surface in the entire Solar System. On average the temperature is around 480 Celsius – enough to melt tin, zinc and lead. The reason for this is the carbon dioxide. Carbon dioxide is a so-called greenhouse gas. It lets in sunlight, but is opaque to the infrared radiation that comprises heat. Venus has become a heat trap.

Venus' clouds are also unlike Earth's. They are composed mainly of concentrated sulphuric acid, often used in batteries. The highest cloud layers float 65 kilometres above the surface – more than four times higher than on Earth. These clouds scoot around at speeds of more than 350 kilometres per hour, driven by the planet's slow rotation. It takes Venus a leisurely 243 days to spin once on its axis. As the sun-facing hemisphere warms up, a high-pressure region develops there, which creates winds that blow clouds around the planet in only four days. At first, astronomers thought that the planet itself was rotating at this speed. But when they glimpsed the surface with radar, they discovered its true rotation period – and more besides.

Beneath the Clouds

Most of Venus' surface consists of rolling planes and lowlands – the 80 per cent of the terrain that has a relief of less than 1 kilometre. Highlands, where the altitude rises more than 2 kilometres above the mean planetary radius, make up the remaining 20 per cent.

Superimposed on this landscape, by far the most common features on Venus are its volcanoes. There are no obvious signs that the planet is still active, but it might well be – and certainly was in the very recent geological past. There are large shield volcanoes like the Hawaiian islands, volcanic domes, extensive lava flows and volcanic craters called calderas. Calderas are created when magma chambers beneath the crust run out of magma (molten rock), and the surface above caves in and collapses. Among the most interesting volcanic features are those that have no analogue on Earth: the pancake domes are an example. They are flat-topped, circular patches of lava that oozed out of the crust, solidified and cracked as they cooled. So-called coronae are also volcanic in nature. These are regions where rising subterranean magma has pushed up the surface and cracked it but not penetrated it – they look somewhat like blisters. Volcanoes are everywhere on Venus, not just concentrated in lines as they are on Earth. This means that Venus' surface is not broken up into plates like Earth's, with volcanoes populating the plate boundaries. Nevertheless the crust does show signs of cracking and stretching that indicates some lateral motion – tectonic activity.

Venus also has its share of impact craters, but they number less than 1000 in total and so are not as common as on Mercury. The reason is that the planet's surface is geologically young. Volcanism has destroyed all but the earliest craters – none from the heavy bombardment has survived to this day. Those craters that exist now were formed only in the last 10 per cent of the planet's history. Also, most of these craters are large: none is smaller than 3 kilometres, and few are smaller than 25 kilometres. The atmosphere is to blame for this. It is so thick that all but the largest meteorites are either completely destroyed or broken up into smaller bits before they hit the hostile surface.

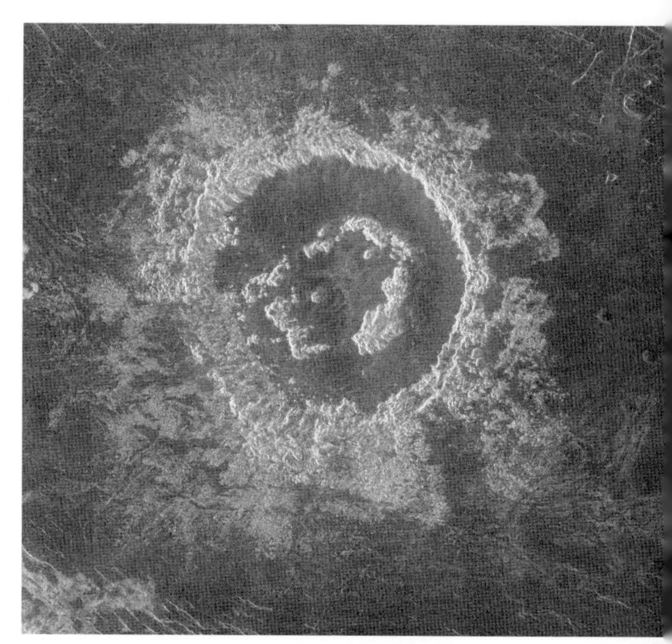

Image above: This radar image shows a curious impact crater on Venus, 50 kilometres across, that has a central mountain ring. The ring was created when the planet's crust rebounded with the force of the impact. The radar-dark floor of the crater indicates that it may have been flooded by lava after it formed. *Courtesy of NASA/JPL/Caltech.*

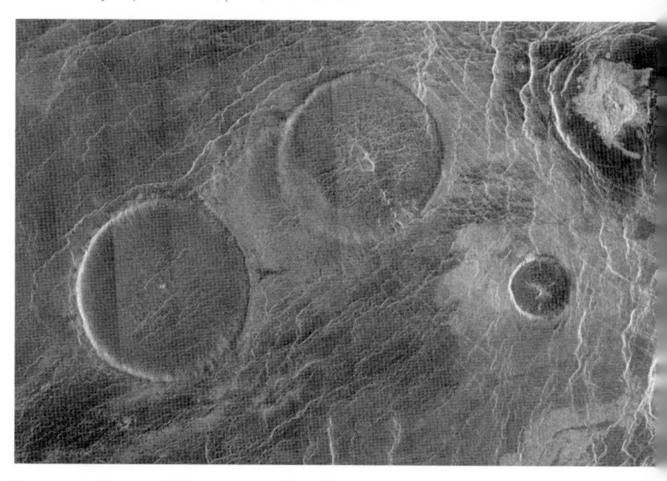

Image above: These circular features on Venus are volcanic domes, where thick lava has oozed out of the surface, spread and solidified. They are 65 kilometres, across but only 1 kilometre high, and have broad, flat tops. Their morphology has led to their informal name, 'pancake domes'. *Courtesy of NASA/JPL/Caltech.*

Evolution of Venus

So why is Venus so hostile? After all, its physical dimensions, mass and density are very comparable to those of our own world. It even formed in a part of the Solar Nebula not too dissimilar from that where Earth formed. And because of this it is quite likely that Venus originally had a lot more water than it does now. Calculations show that the planet may have had enough water to form a global ocean 25 metres deep. This is still a lot less than Earth's own watery oasis, but it is very much greater than the amount of water left on Venus today. Crucially, however, Venus is a lot closer to

thc Sun than Earth is. More than anything else, this proximity is the reason for the planet's gradual evolution into the Hell we see today.

The newly formed Venus was very volcanic, as were all of the new terrestrial planets. Global episodes of volcanism vented huge quantities of water vapour, carbon dioxide and other gases to form the planet's atmosphere. Because Venus is so much closer to the Sun than Earth, it was almost certainly always too hot for any of its water to fall as rain – or at least to pool on the surface. Thus, from the very start, Venus lacked oceans and most likely even seas. This would deal a fatal blow to the planet's prospects for hospitability, because carbon dioxide is soluble in liquid water. On Earth, where it was cool enough for water to pool on the surface, this poisonous gas got dissolved in our oceans, and was thus removed from the atmosphere. But because Venus never had any substantial oceans, if any, it was unable to cleanse its atmosphere of the carbon dioxide vented by volcanoes. So the gas remained in the atmosphere in great quantities, trapping the planet's heat, and the planet grew hotter. The heat in turn baked more and more water vapour and carbon dioxide out of the soil and rocks. The Sun's deadly ultraviolet rays destroyed the water vapour (Venus lacks a protective ozone layer), the leftover carbon dioxide piled up, and the planet grew hotter still… This sad state of affairs is known as a runaway greenhouse effect. Earth was lucky. If it had been just 5 per cent closer to the Sun, it too would have gone the same way, and I would not be here now telling you this.

Aside from Venus' atmosphere, very little is known about the planet's early history. Its oldest impact craters are less than 800 million years old. And on average the surface is 300–500 million years younger even than that. The crust has been completely recycled since it first formed billions of years ago, the handiwork of planet-wide volcanism. But, despite the lack of clues to the past, one other aspect of the planet's history seems certain, and evidence for it comes from the planet's strange, slow rotation. Venus not only spins very leisurely, but it does so backwards compared with most of the other planets. It is hard to see how this state of affairs could have arisen naturally at birth. More than likely, the unlucky world was hit very early in its formation by a gigantic planetoid of comparable size, and knocked over nearly 180 degrees.

Both Mercury and Venus have suffered enormous impacts. Soon it will become apparent that quite a few other planets are the product of chance cosmic encounters, too. In fact the Earth is one of them – as we shall see next.

Image above: This is one of only a handful of images of the Venusian surface, taken by the Russian probe Venera 13 in 1982. The terrain resembles a series of flat plates of basalt, interspersed with soil. *Courtesy of NASA/National Space Science Data Center.*

Image opposite: This global radar map of Venus is a composite of images taken by the Magellan spacecraft. Numerous impact craters are evident, as are several volcanoes. The bright equatorial band is the highland called Aphrodite Terra. *Courtesy of NASA/JPL/Caltech.*

Earth – Goldilocks Planet

Water is abundant throughout the Solar System. But nowhere except on planet Earth are the conditions right for that water to exist as a liquid on the surface. Mars is too cold, its atmosphere too thin. Venus is too hot. But Earth is just right – and for this reason it has been called the Goldilocks planet. Third planet from the Sun and largest of the terrestrial worlds, blue Earth is the gem of the Solar System, almost certainly the only world conducive to complex lifeforms. It has many of the features shared by its terrestrial cousins – volcanoes, impact craters and tectonic activity – and is unique in that its volcanoes are still active. Its single satellite is one of the largest in the Solar System. In fact it is a giant in comparison to its parent planet. It seems to be the outcome of a great cosmic accident – the coalesced debris of a collision that the Earth endured when it had recently formed.

Earth Data

Mass: 5.973×10^{24} kg

Diameter: 12 756 km

Surface gravity: 1.00 gee

Axial tilt: 23.5°

Mean surface temperature: 22 Celsius

Rotation period: 23.93 days

Orbital period: 365.3 days or 1 year

Inclination of orbit to ecliptic: 0.0°

Orbital eccentricity: 0.017

Distance from the Sun: 0.98–1.02 AU

Satellites: 1

Moon Data

Mass: 7.163×10^{22} kg or 0.012 of Earth's

Diameter: 3477 km or 0.27 of Earth's

Surface gravity: 0.17 gee

Mean surface temperature: −42 Celsius

Rotation period: 27.32 days

Orbital period: 27.32 days

Inclination of orbit to Earth equator: 18.3–28.6°

Orbital eccentricity: 0.05

Distance from the Earth: 356 410–406 697 km

Image opposite: From space, the Earth's transparent atmosphere readily betrays a planet unlike any other. Water covers 75 per cent of the surface, either in liquid oceans or frozen at the poles. Its single large satellite, by contrast, is a cratered, airless ball of rock more than one-quarter the diameter of its parent.

Active Planet

While every planet is unique is some way, perhaps none is as different from the others as the Earth is. Three-quarters of its surface is covered in water. Most of this water resides in great oceans, on average about 4 kilometres deep, while the rest of it is locked up at the poles, frozen solid. Like Venus, the Earth's pull is strong enough to retain an atmosphere. But ours is very different. With 78 per cent nitrogen and 21 per cent oxygen, it has almost none of the carbon dioxide that made Venus so inhospitable.

In terms of general composition and overall structure, at least, Earth is not too dissimilar from its terrestrial cousins. Like these other planets, the Earth has a metal-rich core, a rocky mantle and a rocky crust. But there

are some marked differences with our planet's interior. Its core is composed of two parts: an inner and an outer. The inner core – where temperatures may exceed 6000 Celsius – is made of nickel and iron. This is kept solid, despite the heat, by the pressure, all 3.7 million atmospheres of it. The outer, cooler core, under less pressure, has the same composition but is mainly liquid. Both cores account for one-third of the planet's mass, though they occupy only one-eighth of its volume. It is in the core where convective motions generate the Earth's magnetism as they do on Mercury. Above the core is the mantle, composed of minerals such as pyroxene, garnet and olivine. It is mostly solid. However, high pressures and temperatures give the mantle a rubber-like consistency and enable it to flow like a fluid. The uppermost 400 kilometres of the mantle is liquid – a region known as the asthenosphere. Lastly, resting on this pliable bed is a rigid transition zone called the lithosphere, which gradually blends into the Earth's crust. The crust is not a solid block, however, as it is on Venus, Mercury and Mars. It was shattered long ago, by impacts, into several continental plates.

Because the Earth's interior is still hot, its plates cannot remain still. They drift sideways, driven by convection, to create the well-known continental drift. It is this phenomenon that makes the planet so active. Where plates come together, the crust crumples up and forms mountains. And where two plates are being pushed apart, volcanoes fill in the gap with magma that eventually solidifies and forms new crust. Because of Earth's very nature – volcanoes, faults, mountain-building, oceans and rain – the planet has an exceedingly young surface, geologically speaking. Long gone are the scars from impacts endured during the heavy bombardment – and there would have been very many – for most of the crust is only 100 million years old. It is constantly recycled by geological activity and water erosion. Today, only a few impact craters dot the planet – all formed in the recent past.

Dead Moon

The Moon, by contrast, is as different from Earth as it could possibly be. It is too small to retain an atmosphere, and has never supported liquid water. About 84 per cent of its surface consists of a highly cratered, ancient highland. Some of the rocks there – primarily a calcium–oxygen–aluminium compound called anorthosite – are 4300 million years old. The craters indicate that this part of the Moon has changed little since then. The remaining 16 per cent of the surface is taken up by relatively smooth and dark maria (singular mare), Latin for 'seas'. It is the contrast of these dark patches with the brighter highlands that creates the illusion of the Man in the Moon. The maria are not real seas, of course – at least, not of water. They are in fact seas of solidified lava, regions where very fluid magma oozed onto the surface through cracks in the crust and spread out. Thus the

maria are similar to the lowland plains on Mercury and, as on that world, are mostly associated with the large craters called impact basins. Because the maria have relatively few impact craters they are younger than the highlands – but still geologically ancient.

Our Moon is unique among the terrestrial worlds. Of the other rocky planets, Mercury and Venus have no satellites, and those of Mars are puny pebbles, barely kilometres across. Why is Earth any different? The answer, it seems, can be found by looking elsewhere in the Solar System – at Mercury.

Image above: More than half of the Earth's diameter is taken up by a dense metallic core of nickel and iron. The inner core (white) is solid, but the outer core (orange) is molten. On top of this is the mantle, a case of solid rock that nevertheless, because of the high temperature, flows like a plastic. And lastly there is the crust, 10–70 kilometres thick – thinner in relation to the planet's radius than the skin of an apple. The inset shows the Earth compared in size to its Moon.

Formation of the Moon

We have seen that Mercury's high density might be explained if the planet suffered a collision with a gigantic protoplanet as it was approaching full size. In the early 1970s, astronomers put forward a similar suggestion to explain the origin of Earth's large Moon. They call it, quite plainly, the giant impact hypothesis. This scenario is by no means certain. But it does explain a lot of characteristics of the Earth–Moon system and about the bodies in general.

Image above: This image, taken during the Apollo 17 mission, shows the view across Mare Imbrium and towards the 107-kilometre crater Copernicus (top). Imbrium is just one of several 'seas' on the Moon where lava flooded to the surface and solidified, burying more ancient features. Copernicus, meanwhile, is situated in the older highlands. *Courtesy of NASA/National Space Science Data Center.*

Image opposite: The currently accepted theory for the origin of the Moon involves an off-centre impact with the Earth just after the planet had finished accreting. The impactor was a body at least as big as Mars. Most of the debris from the collision fell back to the young Earth, but some of it remained in orbit where it very quickly accreted to form the Moon.

So the theory goes, shortly after the Earth had finished accreting and its crust had started to cool – around 4500 million years ago – the planet was struck a devastating blow. A rogue protoplanet, at least as massive as Mars and possibly three times that, slammed into the primordial Earth in a gigantic off-centre collision. The impact melted much of Earth's crust and mantle and liquefied the impactor almost completely. In the case of Mercury's similar event, most of the debris from the collision eventually fell towards the Sun. This left behind the iron-rich planetary casualty that Mercury is today. But the Earth is more massive than Mercury. Most of the pulverised remnants were captured by Earth's gravity. Some of this material – possibly most of it – eventually fell back to the planet. But a large part of it stayed in orbit long enough for it to accrete into larger and larger chunks – just as the planets themselves had been built earlier. This accreting body would eventually become the Moon. At first, the Earth and Moon would have been very close together, only around ten Earth radii apart. Since the event, though, the two bodies have been slowly moving away from each other. Today, 384 400 kilometres separate them – a distance that increases by about 3 centimetres every year. Your fingernails grow by the same amount in the same time.

In many ways, the giant impact – if it really happened – marks the true naissance of the Earth and Moon. If it had not been for the collision, today's Earth would be a bit smaller than it actually is, and lacking a satellite. Since that birth, the Earth and its satellite have followed radically different evolutionary tracks.

Evolution of the Moon

The Moon's history is doubtless much like Mercury's. Its oldest rocks date to about 4300 million years, and so the Moon was molten for its first 200 million years. Denser materials sank, while the lighter materials such as anorthosite were buoyed up.

Once the surface had hardened, it gradually began to accumulate craters – courtesy of heavy bombardment. Some of these impacts formed the gigantic basins such as Imbrium and Orientale, which fractured the crust locally to considerable depths. Magma from the still molten interior found its way through the weakened lunar crust to the surface where it formed the maria. This flooding occurred around 3900–3000 million years ago, in a series of stages rather than all at once. But, by 3 billion years ago, the Moon's interior had cooled to the degree that no more lava would ever reach the surface. With no atmosphere, or water, and with an end to flooding, the surface changed very little in the last 3 billion years. The only events that have occurred since then are the impacts, though at a much lower rate than during the heavy bombardment. This steady rain of rock pulverised the surface rocks into the fine powder, the regolith, that now covers the lunar landscape. And in the meantime the Moon's interior has frozen solid. Like Mercury, it is a geologically dead world.

Image above: Lunar Orbiter 5 view of Hadley Rille on the Moon. Hadley is a V-shaped valley carved long ago by flowing lava, one of only a few features on the Moon that are the direct result of volcanic activity. When Hadley was created it would have had a roof of solidified lava, but this caved in with the many impacts that the Moon has endured. *Courtesy of NASA/National Space Science Data Center.*

Evolution of the Earth

Earth started out much the same way as the Moon – a molten ball of rock. While the planet was cooling, gases previously locked up in rocks were released from the surface. This outgassing took place on the early Moon too. But, because the larger Earth has six times more gravity, the freed gases clung to the surface as they did on Venus. They formed a primitive atmosphere.

The commonest gases vented by the early Earth included carbon dioxide, methane, ammonia – and water. At first, the Earth's surface would have been too hot for the freed water to exist as a liquid there. It remained suspended in the atmosphere in vapour form. But, as the planet cooled down, there came a point where its surface temperature dropped below the boiling point of water. At last, the water that fell as rain stayed on the surface – and the first oceans began to appear. This process started quickly, within 100–200 million years of Earth's formation. Comets and asteroids, flung into the inner Solar System by the giant planets, especially Jupiter, may also have brought significant quantities of water to the new-born planet, nurturing its oceans still further. Life appeared in those oceans after about 1 billion years. And it was also at about this time that the Earth's surface – cracked into plates by overwhelming cosmic impacts – began to spread sideways, carried by convective motions in the warm mantle beneath. This created the first mountain chains.

With the appearance of the oceans, the atmosphere began to change. Atmospheric carbon dioxide dissolved in the water, where it chemically combined with other materials to form such minerals as limestone. Slowly, the Earth's oceans cleansed the skies of carbon dioxide – a process that had been unable to occur on Venus' hot, waterless surface. After another 500 million years – 3 billion years ago – the atmosphere consisted primarily of those other materials vented by volcanoes: a noxious cocktail of methane, ammonia and other hydrogen-rich compounds. Back then, there was very little trace of free oxygen or ozone in Earth's skies. Ultraviolet radiation from the Sun steadily attacked the hydrogen-rich gases in the atmosphere and broke them apart into their constituent atoms. Nitrogen was freed from ammonia, carbon was released from the methane, shattered water molecules gave up their oxygen – and the released hydrogen escaped into space. Some of the oxygen combined with the carbon to form carbon dioxide, which again was absorbed by the oceans. Meanwhile, other oxygen atoms combined in threes to form ozone. Slowly, the atmospheric stock of nitrogen grew as oxygen and carbon were consumed, but in time the atmosphere stabilised as the ozone began to shield the planet from the Sun's ultraviolet attack. Since then, the evolution of our atmosphere has been dominated by one major force: life.

About 2 billion years ago, plant activity and photosynthesis all over the planet suddenly bloomed. Photosynthesis absorbs carbon dioxide from air or water, uses sunlight to manufacture planet-nourishing carbohydrates and releases oxygen as a by-product. With the onset of mass photosynthesis, free oxygen began to accumulate in the atmosphere. One billion years ago the oxygen levels were about 10 per cent what they are now. But the amount of free oxygen increased dramatically and reached modern levels about 600 million years ago. At that time there was also a huge proliferation of complex life, the Cambrian explosion. The Earth was at last in full swing.

As far as we know, life is unique in the Solar System – certainly complex life is. But there is one other planet on whose surface liquid water almost certainly existed in the past. Could life have arisen there too? Let's turn now to the fourth planet from the Sun: Mars.

Image below: Virtually all of Earth's impact craters have been wiped from the face of the planet by geological forces. Those that remain are mostly very recent, formed within the last 100 million years. These impacts are testament to the pummelling our planet still receives from time to time – often with devastating effects such as the extinction of the dinosaurs.

Mars – Red Planet

As we head away from the Sun we have one more stop before leaving the realm of the inner planets: the Red Planet, Mars. Only 53 per cent the diameter of Earth, Mars is a midsize terrestrial world, accompanied by two small moons. Impact craters dominate its southern hemisphere, but extensive volcanism has significantly modified the north. Red sand, rich in particles of rusted iron, covers the frigid surface, blown by huge, global storms. Meanwhile, the polar regions sport extensive ice caps – though they are made of frozen carbon dioxide as well as water. Carbon dioxide is also the primary gas in the thin Martian atmosphere, as it is on Venus. But Mars has lost most of its atmosphere, and its water is frozen at the poles or embedded in the ground as permafrost. Now, the Red Planet is a cold, hostile desert.

Mars Data

Mass: 6.42×10^{23} kg or 0.11 of Earth's

Diameter: 6794 km or 0.53 of Earth's

Surface gravity: 0.38 gee

Axial tilt: 25.2°

Mean surface temperature: −23 Celsius

Rotation period: 24.62 hours

Orbital period: 1.88 years

Inclination of orbit to ecliptic: 1.8°

Orbital eccentricity: 0.093

Distance from the Sun: 1.38–1.66 AU

Sunlight strength: 0.36–0.52 of Earth's

Satellites: 2

Largest satellite: Phobos, diameter 27 km

Physical Overview

Image opposite: A view of the Red Planet, showing the Tharsis bulge that is home to the Solar System's largest volcanoes. The great canyon, Valles Marineris, cuts across the bottom half of the picture, and the northern polar cap is readily in view.

Mars, more than any other planet, has long been the subject of much human fascination. Nineteenth-century astronomers perpetuated the vision that Mars was covered in lush vegetation, irrigated by a clever network of canals – evidence, they thought, of intelligent Martians. The planet even has the same axial tilt as the Earth, a very comparable rotation period of 24.62 hours, and dramatic seasons. Thus the romantic view of Martian life persisted well into the twentieth century. But when the first probes reached Mars in the 1960s and 1970s, the truth was finally revealed. Mars is not alive. It is dead, and looks as if it has been that way for a long time. No conclusive evidence for life there, either now or in the past, has ever been found. Instead, the planet in some way resembles parts of Mercury, the Moon and Venus.

Like all these worlds, Mars has roughly two different terrains. The highlands, which dominate the southern regions, are heavily cratered. In the north are rolling lowland plains, which have few craters and are more recent than the highlands. One exception to the general low altitude of the northern territories is the Tharsis rise. This huge bulge rises some 10 kilometres above the mean level of the terrain in the north, and runs 8000 kilometres across the planet. Astronomers suspect that Tharsis was created either when magma welled up under the planet's crust and pushed it skywards, or

Image above: Evidence of liquid water in Mars' past is commonplace on the Red Planet. This image by the Mars Global Surveyor spacecraft shows valleys that were created, at least partially, by running water. *Courtesy NASA/JPL/Malin Space Science Systems.*

when the magma flowed onto the surface itself in multiple episodes and solidified. Possibly both mechanisms were at work. Evidence for this volcanic origin is strong. A thong of massive shield volcanoes crowns Tharsis. Three of them lie in a line, and one – the 600-kilometre-wide Olympus Mons – is off to one side. These volcanoes are enormous, evidence that Mars, like all terrestrial planets except Earth, lacks plate tectonics. The magma beneath the stationary crust simply oozed onto the surface and piled up. On Earth, by contrast, this pile-up cannot happen because the continents never stay in one place. Also, mountains are absent on Mars – further evidence for a stationary crust. But there are some tectonic features. The largest and most awesome is Valles Marineris. This is a truly enormous canyon, south of the Tharsis bulge. It is 8 kilometres deep and 4500 kilometres long – the Earth's Grand Canyon, by comparison, is little more than a scratch. Valles Marineris may have been created with the Tharsis uprising, the equatorial crust literally ripped apart as magma pushed its way up further north.

The lack of plate motion on Mars could mean that its crust is thicker than Earth's. The planet cooled more quickly than Earth, being smaller, and so its crust did not fracture with the impacts from space. Underneath the crust, Mars is similar to Venus. It has a thick, rocky mantle and a metal-rich core. And, like Venus, Mars has almost no detectable magnetic field.

Atmosphere and Climate

Like Earth, Mars has polar caps. As well as water ice, they include significant quantities of frozen carbon dioxide: dry ice. But because Mars' orbit is quite eccentric – its average distance from the Sun of 229 million kilometres (1.52 AU) varies by 11 per cent – the polar caps change dramatically with the seasons.

The planet's closest approach to the Sun brings a swift summer in the southern hemisphere and sees the gradual melting of the southern polar cap. Temperatures in the summer can reach 22 Celsius at southern mid-latitudes, and a sweltering 37 Celsius at the subsolar point – but this is very exceptional. Mars is usually very, very cold. And when furthest from the Sun and moving at its slowest, the planet plunges into a deep and extended freeze, −125 Celsius at the south pole. Carbon dioxide – which is the principal component of the Martian atmosphere – then condenses out in the sky and falls to the ground as snow, and the south polar cap gradually creeps across the surface of the planet to reclaim the ground coverage lost in the summer. The water in the polar caps, however, remains frozen at all times; Mars never gets warm enough at the poles to melt water. And even if the water did melt, it would evaporate straight away because Mars' atmosphere

Image opposite: Waterworld – but this is no Earth. This is how Mars may have looked some 2500–3500 million years ago when it was warm enough to support liquid water. Its volcanoes were active at this time too. The Martian moon Phobos hangs in the sky.

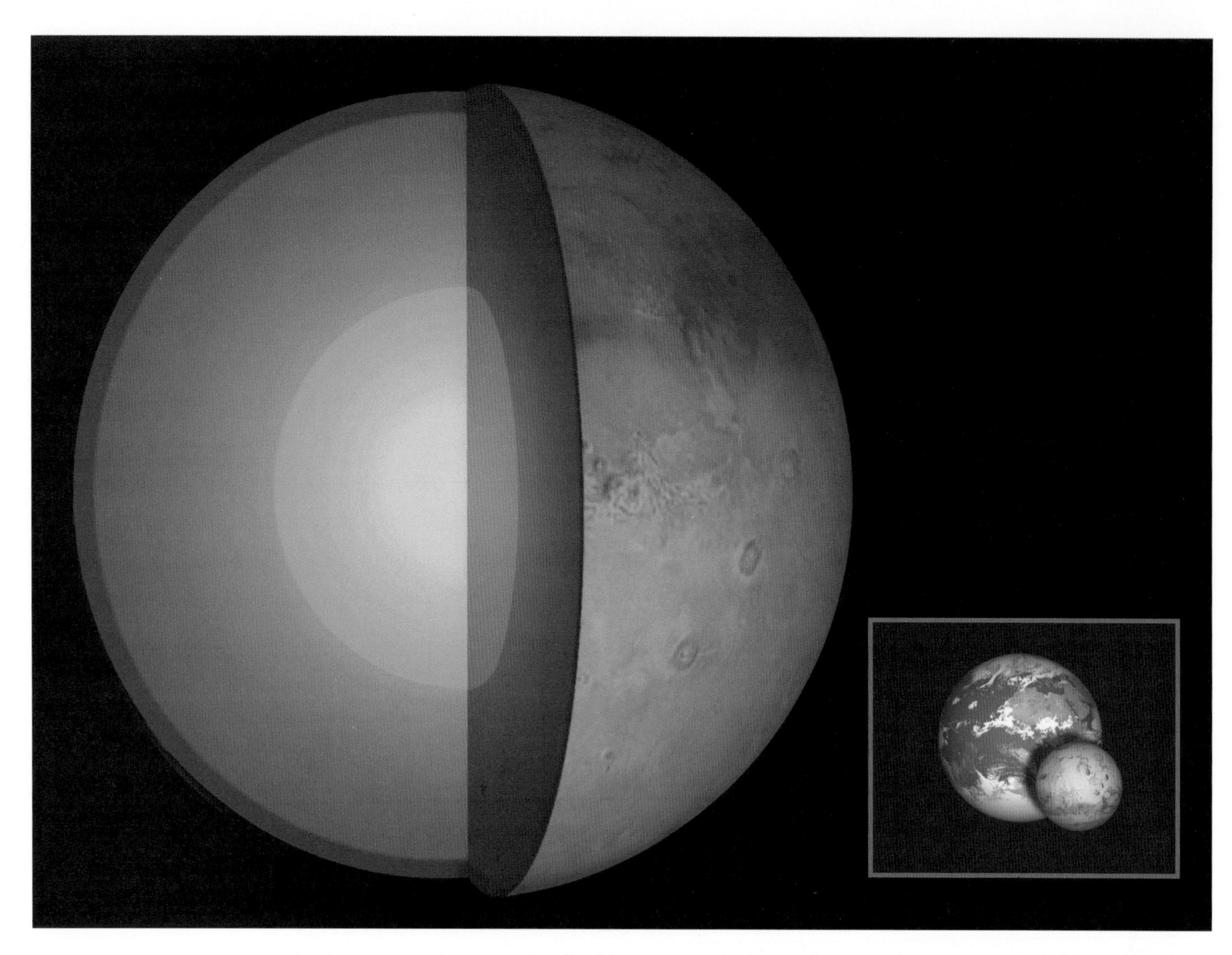

Image above: Mars' interior looks somewhat like Venus'. The large core of iron and iron sulphide may be partially molten. The mantle is similar in density to Earth's and composed of olivine. The crust is thicker than on Earth, though: about 120 kilometres deep. The inset shows Mars on the same scale as the Earth.

is exceedingly thin. It is 100 times thinner than our atmosphere, the pressure at its surface equivalent to that at an altitude five times higher than Everest on Earth. Without a spacesuit, you'd last less than a minute in the cold, dry semi-vacuum that hugs the hostile Martian surface.

Evolution of Mars

Mars' atmosphere was not always this thin, though. There is lots of evidence that liquid water once flowed on Mars, cutting canyons and riverbeds as it does on Earth today. The Red Planet's atmosphere must once have been much denser. The reasons for its unfortunate transformation are many. But two of the main culprits are the planet's diminutive size and its lack of magnetism.

Mars is very small compared with its neighbours Earth and Venus, which are 9.3 and 7.6 times more massive respectively than the Red Planet. Why it ended up with so little mass may have something to do with its position in the Solar System. Mars is found near the inside edge of the asteroid belt. The asteroids are leftovers from the planet-building process.

As we shall see, they were unable to form a planet because of the gravitational field of nearby Jupiter, the next planet from the Sun after Mars. It is possible that Jupiter's disruptive influence was felt even where Mars was forming. The giant's gravity ejected many planetesimals out of the plane of the Solar System, leaving Mars to mop up the scraps. Still, Mars was massive enough to hold onto the atmosphere that it gradually outgassed as it cooled – but only just. Because the Red Planet is so small, much of its original atmospheric gas has slowly escaped. Impacts would no doubt have heated and stirred up the atmosphere during the heavy bombardment phase. Hot gases have faster-moving molecules than cooler gases, so as the atmosphere warmed up it gradually slipped from Mars' weak gravity and leaked away into space. Moreover, because Mars lacks a magnetic field – and may have done in the past – it has no protection from the steady flow of particles from the Sun, the solar wind. Earth's magnetic field deflects the solar wind. But on Mars the wind brushes up to the planet and gradually strips it of gas – up to 45 000 tonnes are lost every year like this.

Gradually, as Mars' atmosphere grew thinner, it also grew colder because of the reduced greenhouse effect. Its atmospheric water was

Image below: This picture shows the surface of Mars as photographed by the Viking 2 probe in 1976. Boulders bestrew the rust-red landscape. *Courtesy Mary A. Dale-Bannister (Washington University), and NASA/National Space Science Data Center.*

destroyed by the Sun's ultraviolet light. Its surface water either froze solid at the poles or, when the pressure got too low, evaporated and joined the atmosphere where it too was destroyed by ultraviolet light. The rest of Mars' water might have seeped into the soil where it still exists in a layer of permafrost. Liquid water may not have flowed on Mars now for 2500–3500 million years. (New research, however, shows that liquid water could exist in small quantities on Mars in very low-altitude regions.) Its volcanoes have also stopped erupting – or so we think. Today, Mars is cold and hostile, probably inactive, and it has not changed in billions of years. If life does exist there, it is certainly not complex. Perhaps it never was.

Image below: The largest Martian volcano is Olympus Mons. It is 600 kilometres across and stands 26 kilometres above the surrounding terrain. On Earth it would stretch from the Isle of Wight to Scarborough and jut three times higher into the sky than Mount Everest. The steep cliffs around its base are 6 kilometres tall. *Courtesy NASA/National Space Science Data Center.*

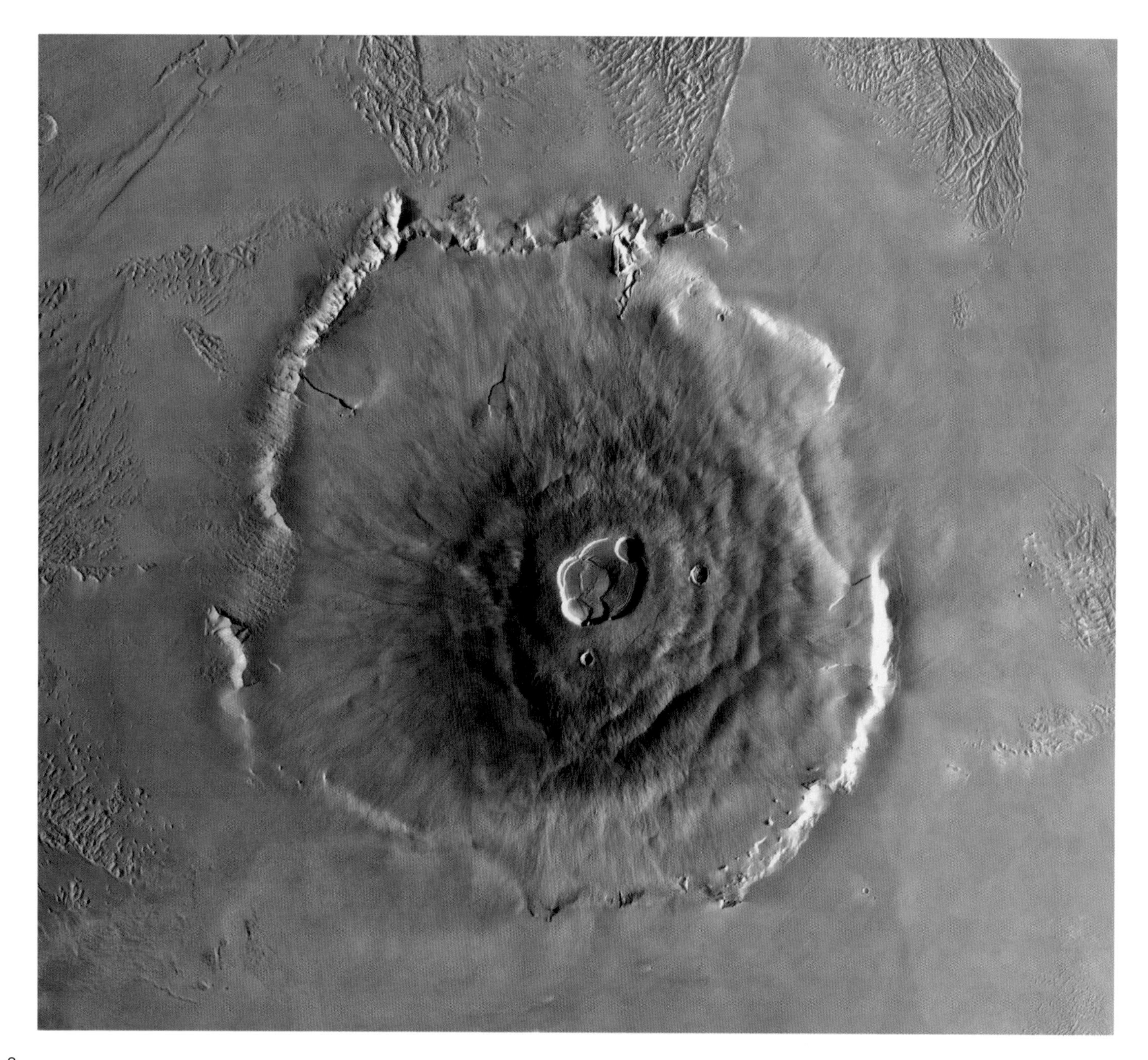

The Dogs of Mars: Phobos and Deimos

Before we leave Mars for the rubble of the asteroid belt, let's meet a couple of the belt's former members: the Martian satellites Phobos and Deimos. Mars' two moons are nothing like ours. They are little more than pebbles in comparison. The larger, Phobos, is only 27 kilometres along its longest axis, about half the size of London. It is irregularly shaped, not spherical, because its gravity is too weak to pull it into a ball. Phobos is the satellite closer to Mars. It orbits 6000 kilometres above the red sands where it swings around the planet in 7.6 hours – less than a Martian day in fact. The other, Deimos, is smaller still, about half the size of its cousin, similarly shaped, and more than 2.5 times further out. Both satellites have exceedingly dark surfaces that reflect just 2 per cent of incident sunlight. They are also heavily cratered.

Phobos and Deimos are not Martian natives but are most likely asteroids. They broke free of their orbits in the belt – assisted by Jupiter's gravity – and headed sunwards where Mars captured them at different times. As to when this happened, though, nobody can tell. The events may date to the beginning of the Solar System, or could have happened much more recently.

Image below: The two Martian moons Deimos (left) and Phobos, as captured by Viking spacecraft, are shown here to the same approximate scale. Both have very dark and cratered surfaces like some asteroids, with which they share a common origin. *Courtesy NASA/National Space Science Data Center.*

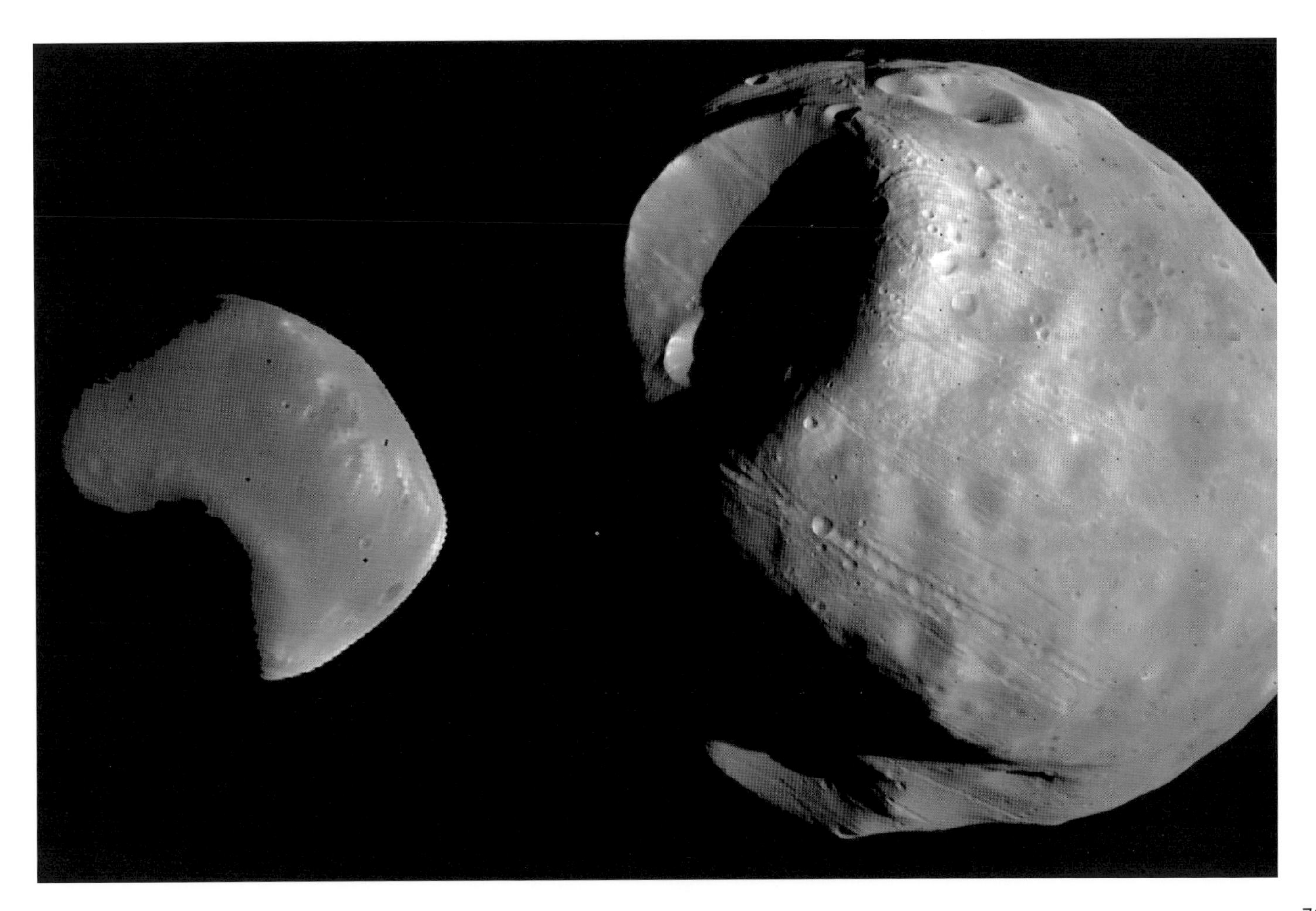

Asteroids – Vermin of the Skies

Between the orbits of Mars and Jupiter is a cosmic dustbin. This is the realm of the asteroids – potato-shaped chunks of rock and metal, heavily cratered, most of them just a few metres across. For much of the time since the first asteroids were discovered, 200 years ago, astronomers have regarded these objects as the 'vermin of the skies'. Often, one of them would pass in front of a more interesting object being photographed, the asteroid's trail as it moved across the field of view marring the result. But lately, astronomers have learned much more about the asteroids – also known as minor planets – than ever before. They are now seen not so much as vermin but as keys to the Solar System's past.

Asteroid Belt Data

Innermost edge: 2.0 AU

Outermost edge: 3.3 AU

Orbital period at innermost edge: 3.2 years

Orbital period at outermost edge: 6.0 years

Known population: >22 000

Estimated population: 500 000 larger than 1.6 km

Estimated total mass: 4.8×10^{21} kg or 0.0008 Earth's

Composition of asteroids: rock and metal

Largest asteroid: Ceres, diameter 900 km

The Asteroid Belt

Image opposite: Even in the main belt the asteroid density is very low. On average, distances of millions of miles separate even the closest members. Most of them, as this artist's impression shows, are lone wanderers.

Since the first asteroid was found in 1801, astronomers have discovered and mapped the orbits of well over 10 000 more. As more and more were found, it became evident that the great majority of them – about 90 to 95 per cent – lie between the orbits of Mars and Jupiter. This region has become known as the asteroid belt. In total, the asteroid belt may harbour a million stony and metal fragments.

When an asteroid orbit is calculated it is numbered sequentially, and the number becomes part of the asteroid's official designation. Thus, the first asteroid to have its orbit mapped is called 1 Ceres. Ceres is easily the largest asteroid at about 900 kilometres across or one-quarter the diameter of the Moon. In fact it is the only one that truly deserves the term 'minor planet'. Ceres is more than twice the diameter of the next largest asteroid, and contributes about one-third of the total mass of all the asteroids combined. Fewer than 20 are larger than 250 kilometres, and most (of those so far found at least) are only about the size of a house or a car. Only the largest asteroids, bigger than about 150 kilometres, are expected to be round. The rest are too puny for their gravities to pull them into spherical shapes. Close up pictures of some asteroids, taken during recent spacecraft missions, confirm this. They also reveal heavily cratered and fractured surfaces, and it seems likely that all of today's asteroids are the product of collisions. With so many asteroids in the belt, close encounters are no doubt relatively frequent. But the romantic image of a vast field densely strewn with madly spinning boulders, popularised by science-fiction films, is very inaccurate. On average, great distances of tens of millions of miles

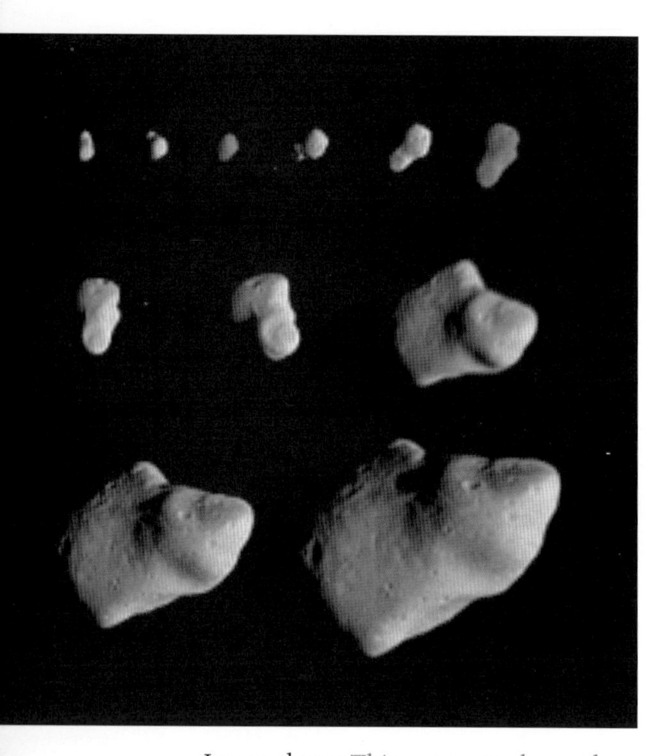

Image above: This sequence shows the asteroid 951 Gaspra as imaged by the Galileo spacecraft in 1991. Gaspra, 17 kilometres long, grew steadily larger as the spacecraft sailed past on its way to Jupiter. The images, the closest taken from 16 000 kilometres, reveal a battered surface. *Courtesy of NASA/JPL/Caltech.*

Image opposite: Some asteroids pass so close to the Sun that they must glow red-hot as they do so. Here, one such asteroid, Icarus, is seen in silhouette against the Sun's torrid disc.

separate each minor planet, and their spin periods can stretch to weeks – though 10–20 hours is more usual.

Asteroids generally orbit the Sun in more elongated or eccentric orbits than those of the planets. Moreover, asteroid orbits are fairly inclined to the ecliptic, the general plane in which the planets encircle the Sun. On average their orbits form an angle of about 10 degrees relative to the ecliptic, but some of them have inclinations as high as 30 degrees. With their high inclinations, the asteroids actually populate a three-dimensional region of space – more a donut than a belt.

The belt extends from roughly 2 to 3.3 AU. Those closer to Mars take about 3 years to orbit, and the most distant belt asteroids take about twice that. Aside from the differences in orbital timescale across the belt, there is also a marked trend in composition. The innermost asteroids, less than about 2.4 AU from the Sun, are mostly stony. They are designated as S-class. They have generally bright surfaces that reflect up to 15 per cent of incident light. The outermost regions of the belt are generally made up of much darker asteroids. These C-class minor planets have a high carbon content and are exceedingly black – darker than soot or blackboards. The moons of Mars, Phobos and Deimos, are captured C-class asteroids. C-class asteroids are also found in the middle of the belt, from 2.5 to 3 AU, along with a third major type, the metal M-class objects. This trend in composition with distance no doubt reflects the different materials that condensed in the Solar Nebula at different distances.

Other Asteroids

Not all asteroids populate the main belt. Some range much nearer the Earth. The so-called Apollo asteroids, for example, which are mostly all smaller than about 5 kilometres, actually cross the Earth's orbit. The Aten asteroids are another group that also cross paths with the Earth. Together, the Apollo and Aten asteroid groups – each named after a single prototype asteroid – are known as near-Earth objects, or NEOs. They, more than any other asteroids, pose a constant threat to life on our planet. Other asteroids have even smaller orbits that hug the Sun, coming well inside the orbit of Mercury. The famous prototype is 1566 Icarus. At its closest approach to the Sun, Icarus' surface temperature soars to 500 Celsius or more.

Other asteroids are found much further out in the Solar System. The Trojans are a group that share Jupiter's orbit. They remain in the same position relative to Jupiter, 60 degrees ahead of and 60 degrees behind the planet, trapped there as a result of the combined gravity of Jupiter and the Sun. And even further afield, some asteroids roam as far as Uranus.

Image above: Conclusive proof that some asteroids come in pairs, this Galileo image shows 243 Ida and its tiny moon Dactyl. Ida, on the left, is a fairly large asteroid 58 kilometres long, but its companion spans at most only 1 kilometre, about the height of Mount Snowdon in Wales. *Courtesy of NASA/JPL/Caltech.*

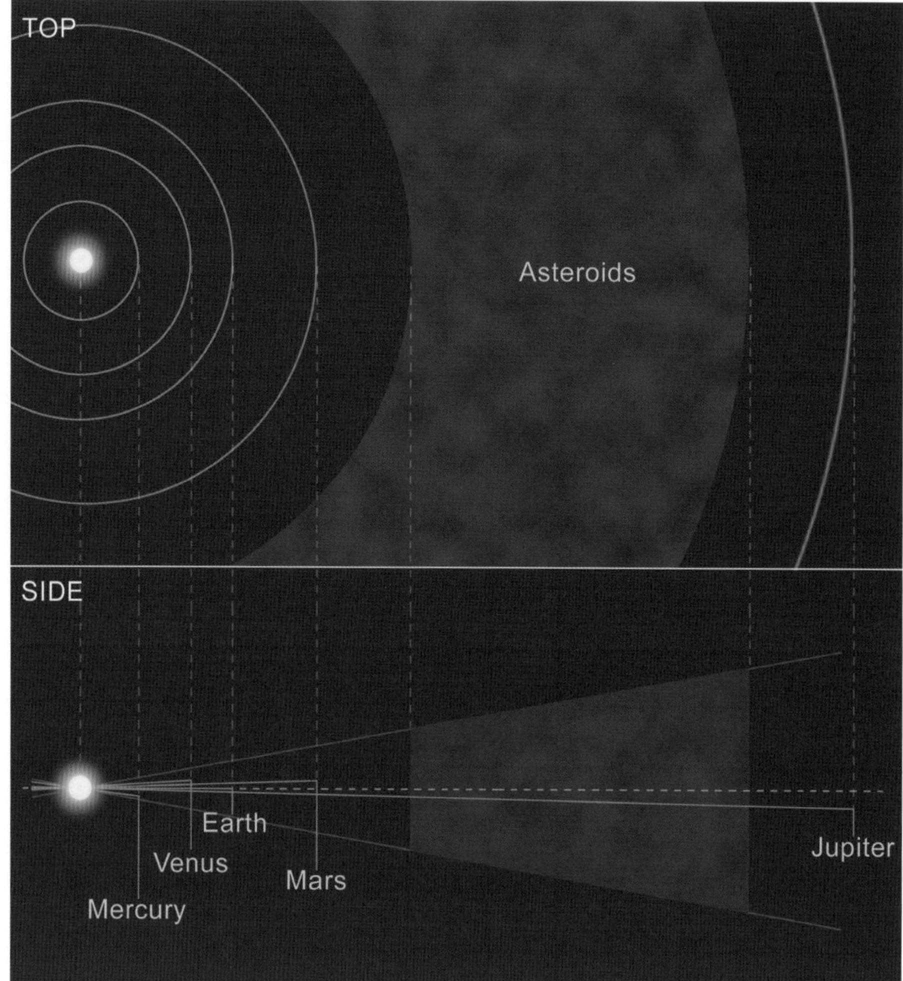

Image right: The asteroid belt is located between the orbits of Mars and Jupiter, and in this image is shown on the same scale as the orbits of the five inner planets. Most asteroids have orbital inclinations up to ±10°, so when the Solar System is seen from the side (bottom) the asteroids delimit a donut-shaped region rather than a flat belt. The inclinations of the planets out to Jupiter are shown for comparison.

History of the Asteroids

Irrespective of where they are found, however, all asteroids share a common origin. Astronomers once considered the asteroids to be the remains of a planet that got pulverised in some giant collision. But the combined mass of all minor planets is estimated to be only about 0.08 per cent of the Earth's mass, or 5 per cent of the Moon's. This is far too little material to make a world even as puny as Pluto. Evidently the asteroids are not the fragments of an exploded planet. It seems more likely that they are the remains – and evidence – of the process that built our Solar System.

Billions of years ago, the first fragments to emerge from the rubble pile that was the Solar Nebula were the planetesimals and then the planetoids. As we have seen, these then gradually lumped together, through gravity, to form the planets as we know them. But the asteroids, it seems –

as well as the comets, which we shall meet later – are testament to the fact that the planet-building process did not entirely finish. One of the reasons for this was the gravitational influence of massive Jupiter. Astronomers are fairly happy with this explanation because, even today, there are apparent zones in the asteroid belt where many members have been thrown out by Jupiter's gravity. These so-called Kirkwood gaps coincide with special orbits that are said to be in resonance with Jupiter's. The ratio of the orbital period of an asteroid in a Kirkwood gap to that of Jupiter is an exact fraction such as 1 : 2 or 2 : 3. This means that an asteroid in a resonance orbit of 2 : 3, say, orbits the Sun three times in exactly the same time it takes Jupiter to go around twice. So, every three orbits of the asteroid – and every two orbits of Jupiter – the asteroid and the planet end up in the same relative positions, and the asteroid gets a regular gravitational kick from Jupiter, each time at the same point in the asteroid's orbit. It is exactly like pushing a child on a swing. If you push at the right time, the swing gains energy and can be made to go very high. In the same way, Jupiter's gravity 'pushes' the asteroids in the Kirkwood gaps at just the right time, and the energy gain makes their orbits wildly unstable. Ultimately they escape from the belt altogether or are flung towards the Sun. This very process enabled Jupiter to depopulate the Solar Nebula where it was forming. In addition, the gravitational perturbations made the planetesimals collide too quickly to stick together, and they broke apart instead. Even today, the asteroids that remain cannot coalesce to form a single planet because of these continuing perturbations.

The largest two asteroids, 1 Ceres and 4 Vesta, are complete planetoids, with ancient surfaces. Other asteroids have obviously been in collision, planetesimals that fragmented. Some even have little moons – most likely bits that were broken off by impacts.

In general, most asteroids have changed little in billions of years. They are among the oldest pieces of the Solar System, and valuable evidence of its traumatic past. They owe their continued existence almost entirely to the presence of Jupiter, our next stop as we move away from the Sun.

Image above: A close up of the surface of 433 Eros – taken by the probe NEAR-Shoemaker from a distance of just 100 kilometres – reveals unprecedented detail. The crater is 5.3-kilometre-wide Psyche, filled with boulders and many smaller craters that formed later. Fractures are also apparent, perhaps the result of another impact elsewhere on the asteroid. Eros measures 36 × 15 × 13 kilometres, and spins once in 5.3 hours. *Courtesy NASA/Johns Hopkins University Applied Physics Laboratory.*

Jupiter – Giant among Giants

Out beyond the rubble of the asteroid belt, like a gargantuan marble in space, we find the fifth planet from the Sun. Giant among giants, Jupiter's size is breathtaking. It spans fully 11 Earths and contains well over 150 times the mass of all the terrestrial planets put together. But it is nothing at all like these smaller worlds. Jupiter's composition is very similar to that of the Sun, mainly hydrogen and helium, and it has no solid surface. We see only its atmosphere. There, clouds encircle the planet in distinct bands, stretched around the globe by rapid rotation, and vast cyclonic storms are commonplace. Because Jupiter is a fluid planet, there are no faults, volcanoes or craters to tell us about the planet's past. But there are clues elsewhere. In attendance around this gas giant is an entourage of at least 16 moons, four of them planet-sized. Their surfaces, at least, have afforded astronomers some understanding of the Jovian system's history.

Jupiter Data

Mass: 1.90×10^{27} kg or 317.7 times Earth's

Diameter: 143 884 km or 11.2 times Earth's

Surface gravity: 2.6 gees

Axial tilt: 3.1°

Mean surface temperature: −150 Celsius

Rotation period: 9.93 hours

Orbital period: 11.9 years

Inclination of orbit to ecliptic: 1.3°

Orbital eccentricity: 0.048

Distance from the Sun: 4.95–5.45 AU

Sunlight strength: 0.034–0.041 of Earth's

Satellites: > 28

Largest satellite: Ganymede, diameter 5262 km

Image opposite: The colourful face of Jupiter as it would appear through a low-power telescope from a distance of about 2 million kilometres. Two of its large moons, Io (left) and Europa, are visible, and several of the frequent lightning storms punctuate the darkness on the planet's night side.

Physical Overview

At roughly 5 AU from the Sun, Jupiter marks the inner boundary of the giant planets. Because of its composition – mostly hydrogen and helium – Jupiter and its ilk have become known as the gas giants. But the name is a misnomer. Though these elements are gaseous under normal atmospheric conditions on Earth, the crushing pressures found inside Jupiter mean that the materials exist there in a liquid form. In reality, Jupiter is an enormous, spinning blob of fluid. It spins so quickly for its size, in less than ten hours, that it is visibly oblate – bulging at the equator.

Deep within its interior, Jupiter has a rocky and icy core. Recall from Part 2 that this was the original giant planetesimal that, after its early appearance in the Solar Nebula, began to suck in the material that made Jupiter so massive. The core is most likely about 20 times the mass of the Earth, and may be separated into two distinct parts: a rocky centre and an icy exterior. However, it is unlikely, given the core's extreme conditions, that its constituents look much like their Earthly counterparts. The core pressure is at least 50 million atmospheres, and the temperature is five times that on the photosphere of the Sun. Meanwhile, above the core, the rest of the planet is virtually entirely liquid. Most of it is hydrogen in a form known as liquid metallic hydrogen. It is under so much pressure that

the hydrogen is split into its protons and electrons and conducts electricity just like a metal – hence the name. Convective motions within the liquid metallic hydrogen, driven by the hot core, are responsible for Jupiter's magnetic field. It is ten times stronger than the Earth's. Above the metallic hydrogen is a smaller shell of ordinary liquid hydrogen. It is under lower pressure and so, though liquid, does not conduct electricity well. It is easy to think that this liquid must have a surface. But this is not the case. There is no sharp boundary separating it from the atmosphere. Instead, the pressure within the liquid hydrogen gradually drops with altitude until, within 1000 kilometres of Jupiter's outer edge, the interior blends imperceptibly into the planet's gaseous shroud – the visible face we see.

The atmosphere of Jupiter is totally different from those of the terrestrials: a different composition, different weather patterns – and a radically different appearance. Its brightly coloured clouds encircle the planet in bands, dragged around by the rapid rotation. The dark bands are known as belts, and the lighter ones are called zones. Probe measurements showed that the zones are higher than the belts in the atmosphere. They are high-pressure regions where gases are constantly rising. In the low-pressure belts, by contrast, the gases are always sinking. Nobody knows how these patterns are maintained. Meanwhile, superimposed on the zones and belts are circular storms, somewhat like hurricanes on Earth. The most famous is the Great Red Spot. It sits about 8 kilometres above the cloud tops and so is a region of high-pressure gas, swirling around as it rises to the top.

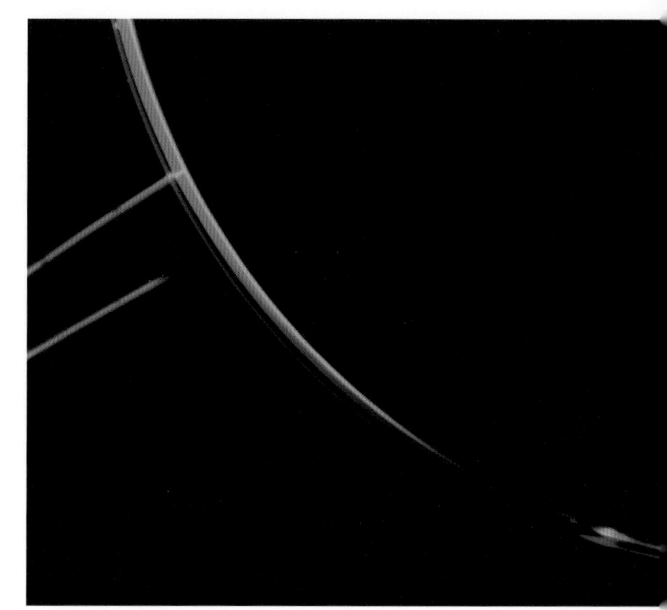

Image above: This image shows Jupiter's thin rings as photographed by the Voyager 2 probe from within the planet's shadow. The rings, almost edge on, are shown in orange. The bottom one appears incomplete because Jupiter's shadow was falling on part of it. *Courtesy of NASA/JPL/Caltech.*

Image opposite: Jupiter's interior consists of a solid core, several times the mass of the Earth, possibly separated into rock (grey) and ice (white). The core is wrapped in a thick blanket of liquid metallic hydrogen, and on top of this, just underneath the atmosphere, is a thinner layer of ordinary liquid hydrogen. The Earth is shown on the same scale.

Rings and Small Satellites

Moving outward from the planet's cloud tops we next come to its system of rings. But the Jovian rings are nothing like the most famous set, those found around Saturn. Saturn's glorious rings, as we shall see, are bright and made of ice particles – some of them the size of houses or even larger. But Jupiter's accoutrements, no more than 30 kilometres thick, are composed of tiny, rocky fragments only about 10 microns across – comparable in size to smoke particles. They are dark and very transparent. And they can only be seen clearly when lit by sunlight from behind.

Because they are so small and lightweight, Jupiter's ring particles are constantly at the mercy of the planet's magnetic field, the solar wind and the Sun's radiation pressure. All exert forces on the fleeting particles and act to disperse them. The fact that the rings are still there means that their stocks of rocky particles are somehow being maintained. How? Micrometeorite impacts on the surfaces of Jupiter's satellites no doubt chip off fragments that may help populate the rings. Or else the rings could be the remains of entire moons that were shattered in the recent past by tidal forces exerted by Jupiter's enormous gravity. It is likely that both

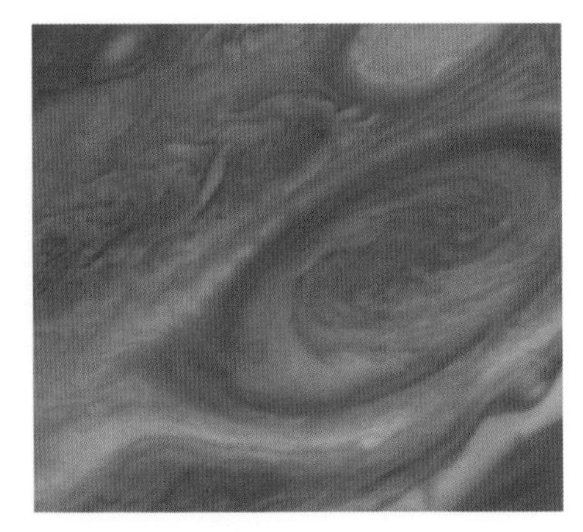

Image above: The Great Red Spot, perhaps coloured by compounds such as phosphine, is a cyclonic storm on Jupiter. It measures 40 000 kilometres across its longest axis – more than three times the diameter of the Earth – and its top protrudes about 8 kilometres above the surrounding clouds. *Courtesy of NASA/JPL/Caltech.*

processes are at work, continually pumping more particles into the rings to replace those swept away. Certainly there are plenty of available satellites. Jupiter has at least 28 of them – and most likely many more remain undiscovered. Most of these satellites are much like asteroids. Like Mars' Phobos and Deimos, they were captured from the nearby asteroid belt.

The Galilean Satellites

Not all Jupiter's satellites are small, though, nor are they all the product of gravitational ensnarement. Four of them are substantial and primordial – that is, they formed alongside Jupiter itself. These are the so-called Galilean satellites, named after the Italian astronomer Galileo Galilei (1564–1642) who discovered them and observed them in the seventeenth century. From Jupiter outwards, they are known as Io, Europa, Ganymede and Callisto.

Io is a rocky world, with little or no ice, a bit larger than our Moon. It is covered in sulphur-rich lava flows and active volcanoes. Around a dozen volcanoes are currently at work on its surface, with the result that Io has the youngest facade in the Solar System. It is only 1 million years old or so, and preserves no trace of the Solar System's early heavy bombardment. Indeed Io has no impact craters whatsoever. They are covered in lava almost as soon as they appear. The reason for Io's activity is its proximity to Jupiter. Tidal forces flex the moon's interior and, like a paperclip bent repeatedly back and forth, it heats up. The heat melts the interior and Io cools down by venting that molten material from its surface in the form of volcanoes. Those volcanoes have furnished Io with a very thin atmosphere. It is little more than a scant shroud of volcanic particles, but it does mean that Io is one of only three or four satellites known to possess a sky. The others are moons of Saturn, Neptune and, possibly, Pluto.

The next Galilean satellite, Europa, is another rocky world, a bit smaller than our Moon. But, in contrast to Io, volcanoes are completely absent on Europa. Being further from Jupiter, Europa no doubt experiences weaker tidal heating. Nevertheless, the moon almost certainly has some internal heat. That this is the case is evident on Europa's surface. Though the moon is mostly rock, it is covered in a bright veneer of water ice, riddled and heavily scored with long cracks, and with less than a dozen impact craters. The lack of craters means than the surface must be young, so the cracks are geologically recent. Astronomers suspect that tidal heating has melted some of the ice beneath the surface to form an underground ocean of liquid water. This keeps the icy crust pliable and frequently fractures it. As a result, Europa's surface is a vast patchwork of ill-fitting jigsaw pieces, giant icebergs, constantly jiggling around – and evidently fast enough to obliterate most impact craters very quickly. We shall see in Part

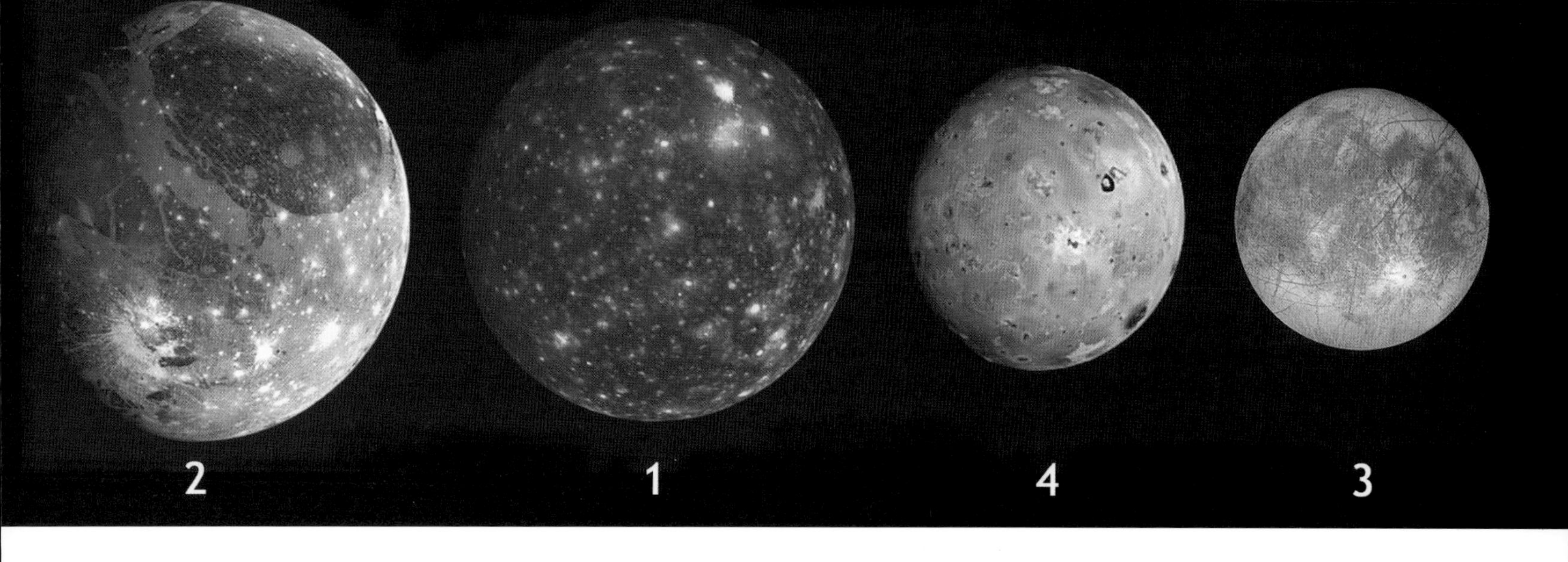

Image above: The Galilean satellites (left to right, Ganymede, Callisto, Io and Europa) are shown here on the same relative scale. Each harbours a unique surface geology. Io and Europa are largely rock, with young surfaces. But, while Io is covered in sulphur, the product of active volcanoes, Europa's surface is a sheet of cracked water ice. Callisto and Ganymede have a greater ice content than the other Galileans, but their surfaces are much older and very cratered. The numbers indicate the moons' positions relative to Jupiter, with 1 (Callisto) being the furthest out. For comparison, our Moon is midway in size between Io and Europa. *Courtesy of NASA/JPL/Caltech.*

4 that Europa's ice might well melt in the far future, when the Sun becomes a red giant. But it the meantime it will remain frozen solid.

Ice also covers the surface of the next moon out, Ganymede – in fact this world is half ice and half rock, the latter forming the core. Ganymede is the largest satellite in the Solar System, bigger than the planet Mercury. Parts of its surface are heavily cratered, and in general the terrain is much darker than Europa's. The dark colour comes from dirt strewn across the surface by aeons of accumulated meteorite impacts – evidence, as well as the craters themselves, that Ganymede's surface is much older than either Io's or Europa's. There are regions that seem to be relatively young, with few craters, but they are still ancient. Covering two-thirds of the surface, this landscape is known as the grooved terrain. It was formed about 1–2 billion years ago, when the ice crust was a bit thinner because of the warmer interior then. Around this time, water from beneath the relatively thin crust flooded the surface – cracked perhaps by tides – and froze. As it did so it filled in the cracks and widened them to create the relatively bright grooves that give this terrain its name. But Ganymede feels Jupiter's tides even less than Europa does. And so, lacking an appreciable amount of internal flexion, it has now frozen solid to considerable depth. Without the volcanoes of Io or the thin surface of Europa, little has modified Ganymede's face in a billion years. It remains old and battered.

Lastly, Callisto, also an ice world, has an even older and darker surface. Tidal forces have affected this world least of all, and no cracking seems to have occurred. Evidently Callisto froze very quickly after it formed. It is more heavily cratered than any other planetary body, even Mercury. This is almost certainly because of Jupiter, whose powerful gravity sucks in comets and asteroids, which then collide either with the planet or its moons. Ganymede, being nearer to Jupiter than Callisto, ought to have more craters, but some of them were no doubt wiped out when the grooved terrain was laid down. If Io and Europa had not eradiated their craters, they would be even more scarred today than Ganymede and Callisto.

History of the Jovian System

Image opposite: Night time on Europa. A flight over the surface reveals a cracked landscape, evidence that an underground ocean of liquid water is constantly modifying the terrain. The Sun is behind and below the horizon, but in the sky hang Io and Jupiter, the latter casting a ghostly yellow light on Europa's fractured face.

Because Jupiter has no surface – and therefore no surface geology – it is very difficult to see how the planet might have changed in the billions of years since it formed. It is probable, however, that the planet has evolved very little. Jupiter's escape velocity is so high that it has retained virtually all of the material that went into its making – even the lightest and swiftest gas, hydrogen. Thus Jupiter's composition is essentially primordial, a chunk of the original Solar Nebula. When the planet was first formed, it would have been a lot hotter than now, and even glowing. Gradually it cooled down; molecules began to form from its gases. Thus Jupiter is quite a bit cooler and a little smaller – because of continued gravitational contraction – than when it stopped accreting. Other than that, little else has changed – or if it has, we cannot easily see the evidence.

On the other hand, the Galilean satellites reveal much about the history of the Jovian system. We saw in Part 2 that these large satellites accreted from a disc that surrounded the newly forming Jupiter, in much the same way that the planets themselves grew in the disc around the Sun. Evidence for this scenario is strong. The Galilean moons all orbit in roughly the same plane – that which defined the original disc, long since gone. They also orbit in the same direction. And there is more evidence in the moons' general compositions and densities. The innermost two, Io and Europa, have high densities. They are rocky, with little ice. Ganymede, further out, has a density between those of ice and rock, and contains equal amounts of each. And Callisto has an even lower density. All of this is to be expected if the moons formed in a disc. Closer in, where it was warm, ices could not condense, and the moons there formed from rock. But further out, where it was cold enough for ices to condense, the moons formed from equal amounts of ice and rock.

Today, the surface of Callisto is our best indicator of the early history of the Solar System near Jupiter. It is much the same story as nearer the Sun, with impact craters betraying the handiwork of the heavy bombardment. Here, though, the bombardment was even more torrential, as more comets and asteroids than usual succumbed to a sticky end, netted by Jupiter's powerful gravity. Since the end of the heavy bombardment, Callisto has changed very little. Ganymede, as we saw, experienced flooding by water more recently. But Europa is still warmed inside from tidal heating. It has an active surface that betrays little about its past, and volcanic Io is the same.

Saturn – Lord of the Rings

Now we are truly entering the depths of interplanetary space. Twice as far from the Sun as Jupiter we encounter the second largest planet. This is Saturn, Lord of the Rings. Saturn is like Jupiter in many ways. The two planets have similar sizes, they share almost identical chemical compositions and their interiors are made of the same layers, though in different proportions. And, like Jupiter, Saturn spins very quickly, its clouds stretched into bands parallel to the planet's equator. But it is the incredible rings that make Saturn so special. They are very bright, made of countless chunks of tumbling ice, and span a volume of space that would stretch more than a third the distance from the Earth to the Moon. Saturn also has a large number of its own moons, including the second-biggest satellite in the Solar System, Titan.

Saturn Data

Mass: 5.69×10^{26} kg or 95.2 times Earth's

Equatorial diameter: 120 536 km or 9.5 times Earth's

Surface gravity: 1.16 gees

Axial tilt: 26.7°

Mean surface temperature: −80 Celsius

Rotation period: 10.23 hours

Orbital period: 26.5 years

Inclination of orbit to ecliptic: 2.5°

Orbital eccentricity: 0.056

Distance from the Sun: 9.00–10.07 AU

Sunlight strength: 0.0099–0.012 of Earth's

Satellites: >30

Largest satellite: Titan, diameter 5150 km

Physical Overview

Image opposite: Saturn is a low-contrast Jupiter, spinning so quickly that it bulges visibly at the equator. Its bright rings are made of chunks of ice, encircling the planet in its equatorial plane. Note the large gap near the outer edge of the rings. It is a region of low particle density known as Cassini's division.

The second largest planet is in many ways a small, low-contrast version of Jupiter, with about one-third of its larger cousin's mass and 84 per cent of its diameter. Like Jupiter, Saturn is a gas giant: a ball of hydrogen and helium in a liquid state. But, because of the comparatively low overall mass of the planet, Saturn's materials are not so compressed. Inside, Saturn hides a dense core of ice and rock, similar to but less massive than Jupiter's. It also has a stock of liquid metallic hydrogen in its interior. But, again because of the relatively low pressure, this region is not as extensive as it is in Jupiter. A thick shell of normal liquid hydrogen tops the liquid metallic hydrogen. And, as on Jupiter, this liquid interior gradually blends into Saturn's gaseous atmosphere.

This hydrogen–helium atmosphere is in some way like Jupiter's. It has dark belts of sinking, low-pressure gas, and brighter zones of rising high-pressure regions. As on Jupiter these belts and zones are stretched around the planet because of the rapid rotation. Moreover, enhanced images of Saturn show that its atmosphere shares some of the other features found on Jupiter, such as the oval storm systems, though on smaller scales. And yet, despite these similarities, Saturn's face is very bland compared with Jupiter's vibrant variegation. This is partly because of Saturn's placement in the Solar System. Saturn is twice as far from the Sun, so its atmosphere receives about one-quarter of the solar energy per unit area that Jupiter does.

Its gases cool quickly as they rise up through the frigid atmosphere, and cloud-forming condensation thus occurs relatively low down where the clouds are shielded from sunlight. In addition, Saturn's outermost atmosphere contains a layer of methane haze. This obscures the clouds below even more and makes the planet's atmosphere paler still. Overall, Saturn is a pallid Jupiter. But what it lacks in atmospheric colouration it more than makes up for in its ring system.

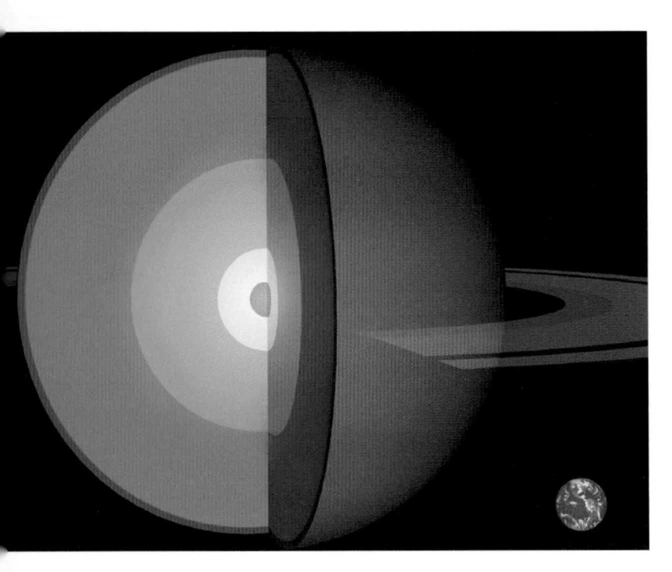

Image above: Inside, Saturn is very much like Jupiter. But the three components – rock/ice core, liquid metallic hydrogen and liquid hydrogen – are in different proportions. Saturn's relatively low internal pressure means that it has less liquid metallic hydrogen than Jupiter, and subsequently a reduced magnetism. The Earth is shown on the same scale.

Saturn's Rings

Like the rings of Jupiter, Saturn's are made up of billions of fragments on independent orbits. But Saturn's rings are very bright and extensive compared with Jupiter's, and this is because the ring particles are different. Saturn's rings are bright because the fragments are icy. The particles are also much larger than the dust-sized, rocky motes in Jupiter's rings, ranging in size from sand-grains up to kilometre-scale, flying mountains.

Saturn's rings are much more complex than was originally thought. Earth-bound observers identified three rings and labelled them A, B and C. Astronomers expected that these individual rings would appear smooth in close-up photographs taken by the Voyager 1 probe. But these close-ups instead revealed that the large rings are themselves made up of thousands of thinner ringlets. Evidently, therefore, the particles are not uniformly distributed. Some of the rings are relatively dark. They no doubt contain darker or smaller particles or are less dense. And in some places there are gaps where virtually no rings are seen at all. The most celebrated of these is the broad Cassini division, roughly 4700 kilometres wide. The Cassini division and other gaps in Saturn's rings are somewhat like the Kirkwood gaps. Recall that these are those regions in the asteroid belt where there are far fewer asteroids than usual. In the same way that Jupiter's gravity sweeps these gaps in the asteroid belt, so the gravity of some of Saturn's moons are responsible for the Cassini and other divisions. Similarly, the gravitational effects of small moons called shepherd satellites are responsible for keeping the individual ringlets so narrow. And Saturn has many moons to do this job.

Image opposite: Seen up close, Saturn's rings are resolved into an expansive system of bright, icy particles, each on an independent orbit about the planet. The particles vary in size, typically from peas up to footballs. But the largest of all might be several kilometres across. The rings are extremely thin, with the relative dimensions of a piece of paper the size of a football pitch.

Titan and Other Satellites

So far, astronomers have found 30 moons around Saturn – more than any other planet – the latest 12 added as recently as 2000 and 2001. Roughly speaking, the satellites fall into three size classes. Titan, easily the largest, is out there by itself, 5150 kilometres across. Much smaller than Titan, meanwhile, are the next six moons, Iapetus, Rhea, Dione, Tethys, Enceladus and Mimas, in order of decreasing size. These form the second class of satellites, are essentially spherical, and have diameters of 398–1440 kilometres. Lastly, the smallest moons are all irregularly shaped and between 13 kilometres and 370 kilometres across. Virtually all of these moons, even the very small ones, contain significant quantities of ice. Titan is half ice and half rock, but for the others the ratio is about 60 : 40 or 70 : 30 ice to rock.

Titan's name is very fitting, for it is a monster satellite. It is the second largest known moon, after Jupiter's Ganymede, and like that world is larger than the planet Mercury. Titan has a remarkably thick atmosphere – even denser than Earth's – which like our atmosphere consists mainly of nitrogen. But Titan is nothing like the Earth. For a start it is exceptionally cold, with a surface temperature of around −180 Celsius. In fact it is because of this frigidity that Titan has an atmosphere at all. At these temperatures, gases move so slowly that they are unable to escape Titan's relatively feeble gravity – around 14 per cent as strong as ours. Instead, the gases cling to the surface in a thick shroud 1.6 times denser than Earth's atmosphere at sea level. That surface, meanwhile, is unlike any other in the Solar System. Planetary scientists have detected various hydrocarbons, and they suspect that parts of Titan are covered in liquid seas or oceans. But they are not oceans of water – it's far too cold for that. Instead, Titan's oceans are made of liquid ethane, methane and nitrogen, possibly enveloping the surface – up to 1 kilometre deep in parts – in a tarry goo. Sadly, nobody has yet glimpsed Titan's surface because its clouds are as impenetrable as Venus'. We shall have to wait until the Huygens probe arrives at this world in 2004 and descends through its dark, orange skies.

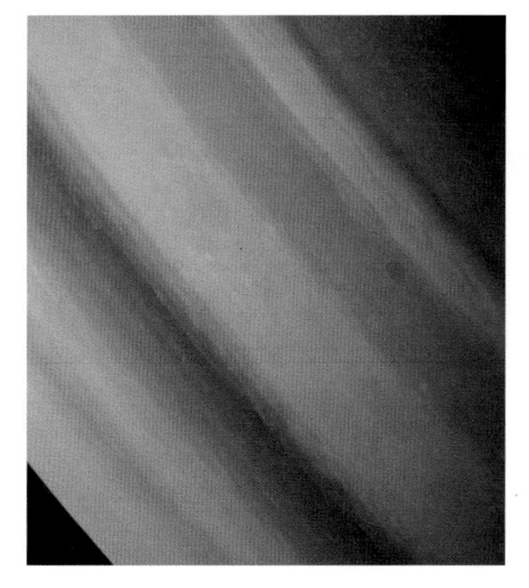

Image above: Saturn's atmosphere, like Jupiter, is banded – the product of rapid rotation. But the extreme cold means that the clouds form lower down where they are less distinct, and a methane haze dulls the colours and makes the atmosphere less vibrant than Jupiter's. This close-up image, taken by Voyager 1 in 1980, reveals two dark oval storms (right), each about 10 000 kilometres across. *Courtesy of NASA/JPL/Caltech.*

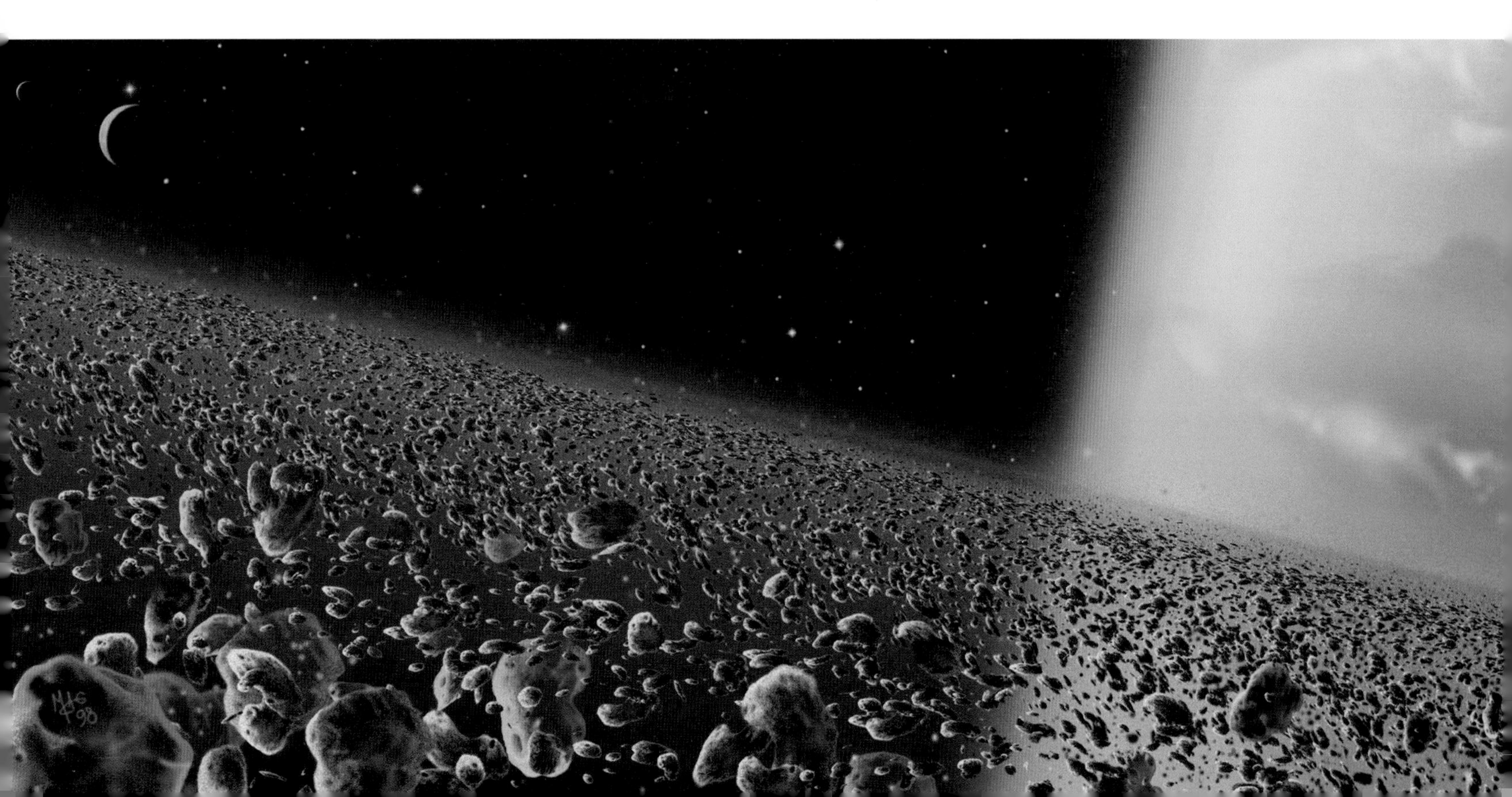

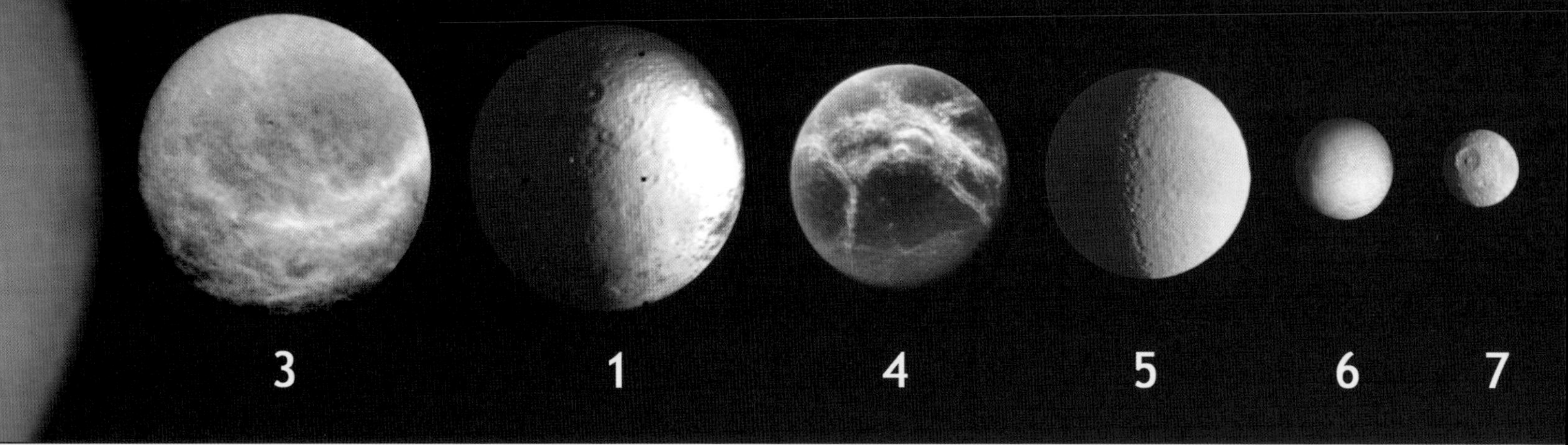

Image above: Like Jupiter, Saturn is surrounded by a large system of varied satellites. This composite shows the seven largest on the same scale, from left: Titan, Rhea, Iapetus, Dione, Tethys, Enceladus and Mimas. The numbers indicate the moons' positions relative to Saturn, with 1 (Iapetus) being the furthest out. For comparison, our Moon is about 67 per cent the size of Titan. *Courtesy of NASA/JPL/Caltech/National Space Science Data Center.*

Image opposite: In late 2004, the hardy little Huygens probe will descend through the thick atmosphere of Titan and provide astronomers with their first ever glimpse beneath this world's permanent cloud blanket. Titan, largest moon of Saturn, is one of only two or three places in the Solar System where astronomers think life might, just might, exist when the Sun grows warmer.

History of the Saturnian System

We have seen already that Saturn formed in the same way that Jupiter did. A large icy planetesimal appeared first, which drew in gas from the Solar Nebula to create a disc around itself. Saturn grew at the centre of the disc. Since its birth, like Jupiter, it has been cooling down and contracting, and even now has a hot interior. However, the rings could not have been formed alongside the planet itself, for the heat would have evaporated them. And in any case, dynamical studies show that the rings are expected to last, at most, only one-tenth the age of the Solar System. Instead, Saturn's icy rings are the remains of large comets or moons that were broken up by the planet's gravity a few hundred million years ago – different in origin to Jupiter's rings.

Further out in the Saturn disc the material presumably lumped together to form many of the satellites. Most of the moons orbit the planet close to its equatorial plane and in the same direction – facts consistent with their formation in a disc. Yet there are irregularities with the Saturnian system that do not conform to a discal birth. The most important discrepancy is that the moons do not seem to be related to each other. Recall that the Galilean satellites have densities that decrease away from the planet Jupiter. This is what we expect: rocky worlds emerge close in where it is hot, and lightweight icy bodies lump together further out. The trouble is that all of the Saturnian satellites are very icy, and they do not show this pattern of decreasing density with distance from the planet. Also, only one of Saturn's satellites, Titan, is anywhere as near as large as Jupiter's big four. The rest are puny. The answer could be that many of the moons have suffered major impacts since their formation – impacts so devastating that the moons' surfaces now bear little testament to the past. The other difference compared with Jupiter is that even Saturn's smallest moons are icy. They are captured icy planetesimals, rather than the asteroids that Jupiter – being close to the asteroid belt – found it easier to net.

Uranus – World on its Side

Once beyond Saturn, we must again journey nearly twice as far from the Sun to get to the next planet. There, 19 times further out than the Earth, we find Uranus – pale blue-green, remarkably featureless, and bitterly cold. Uranus is a mid-sized giant, roughly half the size of Saturn, a fluid blob of slushy ice with smaller quantities of rock and gas. Strangely, this planet rotates on its side, its spin axis lying almost in the plane in which it orbits. Its large array of satellites and rings, encircling the planet's equator, are thus similarly inclined, their orbits virtually at right angles to the Solar System. It is almost as if a gigantic collision knocked Uranus sideways in a game of interplanetary billiards. Indeed, just such an event is astronomers' favoured explanation for Uranus' weird orientation. And it isn't just Uranus that has suffered. The surfaces of some of its moons paint a similarly brutal history throughout the Uranian system.

Uranus Data

Mass: 8.68×10^{25} kg or 14.5 times Earth's

Equatorial diameter: 51 118 km or 4.0 times Earth's

Surface gravity: 1.17 gees

Axial tilt: 97.9°

Mean surface temperature: −214 Celsius

Rotation period: 17.23 hours

Orbital period: 84.0 years

Inclination of orbit to ecliptic: 0.8°

Orbital eccentricity: 0.046

Distance from the Sun: 18.28–20.08 AU

Sunlight strength: 0.0024–0.0030 of Earth's

Satellites: > 21

Largest satellite: Titania, diameter 1600 km

Physical Overview

Image opposite: Uranus is a world on its side, its rotation axis inclined at 98 degrees to the plane in which it orbits. Summer occurs in a given hemisphere when that part of the planet is pointed almost directly at the Sun. Here we see the planet and its dark, narrow rings facing the Sun like a gargantuan bull's-eye in space.

For almost 200 years, little was known about faraway Uranus. But, in 1986, NASA's probe Voyager 2 arrived at the massive green planet. For the first time, astronomers saw Uranus not as a speck of light visible only to the keenest naked eye; they viewed it as a whole new world.

Uranus is another giant, fully four times larger than the Earth. But it is less than half the size of the gas giants and consequently very different. Uranus is relatively dense, so the materials that make it up must be somewhat heavier than the lightweight hydrogen found throughout Jupiter and Saturn. Most of the planet is almost certainly composed of ices – water, methane and ammonia. Because of the internal pressure and temperature, the ices are not solid. Instead, they surround a suspected rocky core in a deep, slushy ocean that occupies two-thirds of the planet's interior. Thus, like Jupiter and Saturn, Uranus is a fluid world; but it is an ice giant, not a gas giant. Pure hydrogen (not that locked up in water, methane or ammonia) and helium make up only 15 per cent of its mass, compared with 80 per cent in Jupiter. And unlike on that world, where the hydrogen is almost ubiquitous, in Uranus it exists only in the atmosphere or in a comparatively thin 'mantle' region between the icy slush and the atmosphere. The atmosphere, meanwhile, is strikingly bland – a featureless green. The grandiose colours and bands that typify the gas giants are absent. The reason

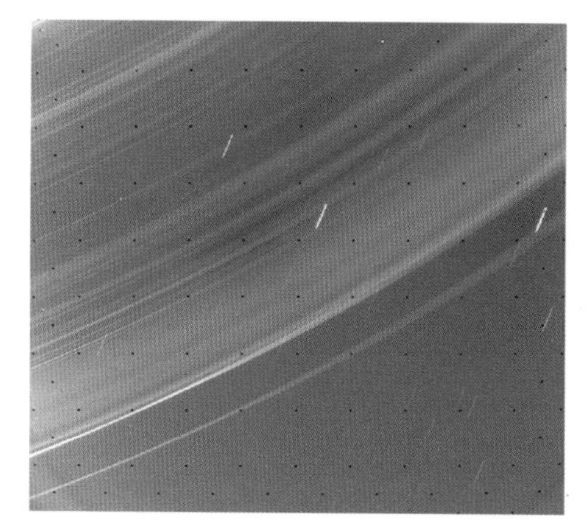

Image above: The Voyager 2 spacecraft returned this view of the Uranian ring system while the probe was in the planet's shadow. Background stars are streaked because of their movement during the length of the exposure (96 seconds) needed to capture the rings properly. *Courtesy NASA/National Space Science Data Center.*

Image right: Uranus has a different interior than Jupiter or Saturn. A slushy outer core (white) of ice-rich compounds including water ice surrounds a rocky inner core (grey). Because of the planet's smaller size, it lacks liquid metallic hydrogen. Instead, the ice is topped by a relatively thin blanket of liquid hydrogen, mixed with other gases, gradually blending into the planet's gaseous atmosphere. The Earth is shown on the same scale.

Image opposite (top): Though Uranus is nearly featureless optically, this false-colour image taken by the Hubble Space Telescope in the infrared region of the spectrum reveals detail. The colours represent different layers of gases in the atmosphere, indistinguishable in optical images. *Courtesy E. Karkoschka and NASA.*

for this is the low temperature. Uranus, twice as far from the Sun as Saturn, is unbearably cold, so frigid that its clouds condense very low down in its atmosphere, where it is warmer. The clouds are so deep that other atmospheric layers hide them. The green colour, meanwhile, comes from a layer of methane high in Uranus' atmosphere. This gas absorbs red light and reflects primarily blue and green.

Perhaps Uranus' oddest aspect is its axial inclination. While the Earth is tilted with respect to its orbital plane, the ecliptic, by 23.5 degrees, Uranus lies at an angle of nearly 98 degrees. During summer in the north, the northern hemisphere is pointed within only 8 degrees of the Sun. The south-polar regions then endure a bitter, sunless night that lasts for almost 21 years. After that time, when the planet has completed one-quarter of its 84-year orbit, its equatorial regions then face the Sun as they do on a 'normal' planet. Then, 21 years later, the north pole is plunged into darkness while the south pole enjoys its long summer – if you can call it that. Only Pluto matches Uranus for these bizarre seasonal variations. And it is because of Uranus' strange tilt that the planet resembles – with its system of rings – a vast target in space.

Ring System

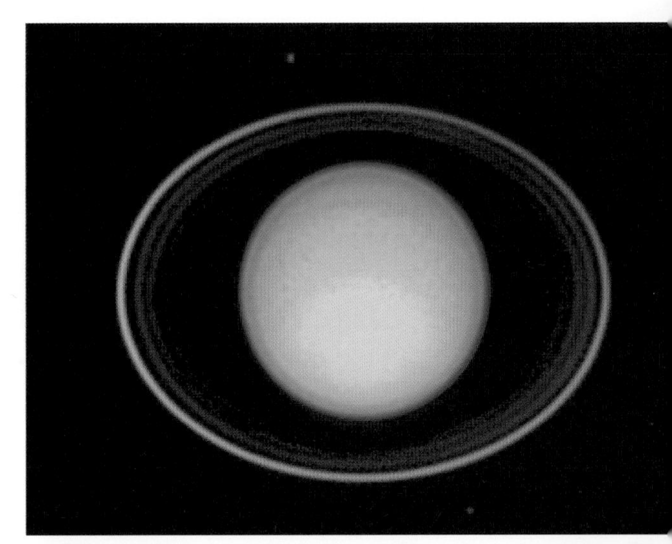

Uranus's rings are different from those of the gas giants, which are in turn different from each other. Its most substantial rings are the so-called 'classical' ones – the nine that were discovered from Earth in 1977. The fragments that make up these rings are typically metre-sized boulders, a little larger in size than the inhabitants of Saturn's great accoutrements. But in stark contrast to the bright, icy particles in Saturn's rings, those that populate the Uranian versions have exceptionally dim surfaces. They reflect only 4–5 per cent of incident sunlight and are thus about as dark as chunks of coal. In addition to and interspersed with the nine main rings, Uranus has a whole range of others, too transparent to be seen from Earth. These are just as dark as the classical rings but they are made up of far smaller particles – dust grains, like those in the rings of Jupiter.

All of these rings, including the classical ones, are extremely narrow. Most are no more than 10 kilometres in radial extent, and even the widest spans only 100 kilometres – 0.2 per cent of the diameter of the planet. They are kept so narrow because Uranus, like Saturn, plays host to a series of so-called shepherd satellites – tiny moons, mere tens of kilometres across, whose gravitational influences herd the ring particles and prevent the rings from spreading out.

Uranian Satellites

Unlike Jupiter, Saturn and – as we shall see – Neptune, Uranus has no very massive satellites. Its largest five measure between just 480 kilometres and 1600 kilometres across, much smaller than the Earth's Moon. From the innermost outwards, they are Oberon, Titania, Umbriel, Ariel and Miranda. Umbriel and Oberon are both heavily cratered, and their surfaces appear to have been flooded by icy 'lava' long ago in the past. Titania and Ariel are cratered too, but their surfaces are also riddled with vast cracks and faults, evidence of past tectonic activity perhaps brought about by tidal heating, as on the Galilean satellites. Lastly, Miranda, the smallest, has very likely the strangest surface in the entire Solar System. Put simply, it looks like a patchwork. Adjacent areas, separated by sharp boundaries, seem to belong to different worlds. One possible explanation for its jumbled appearance is that Miranda suffered a collision so devastating that the moon shattered and re-formed in orbit. Alternatively its appearance could also have been caused by internal melting. Miranda and its four cousins all have fairly low densities, yet they are a bit denser than Saturn's moons. They have a bit more rock than ice. But their surfaces are fairly dark because of dirt spread by aeons of impacts.

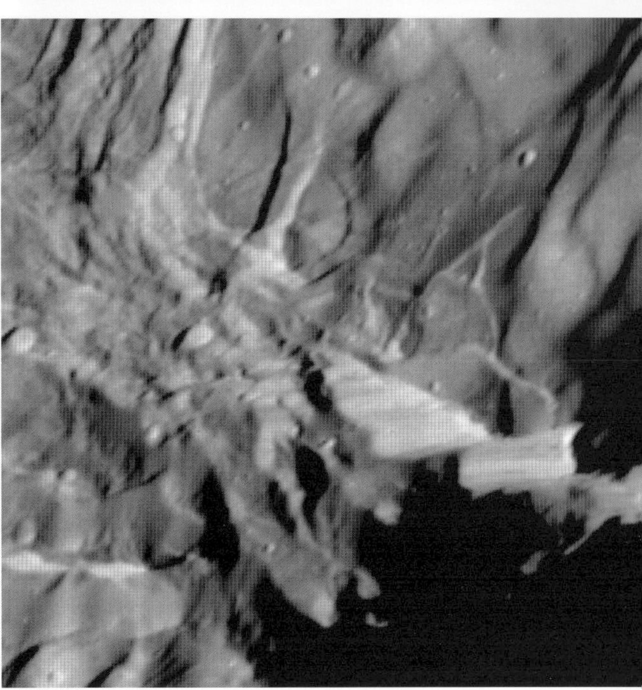

Image above: A close-up of Miranda's bizarre surface reveals numerous ridges and valleys. Most dramatic of all, though, is the great fault at bottom right. Its 45-degree slopes are estimated to be 5 kilometres high. *Courtesy of NASA/JPL/Caltech.*

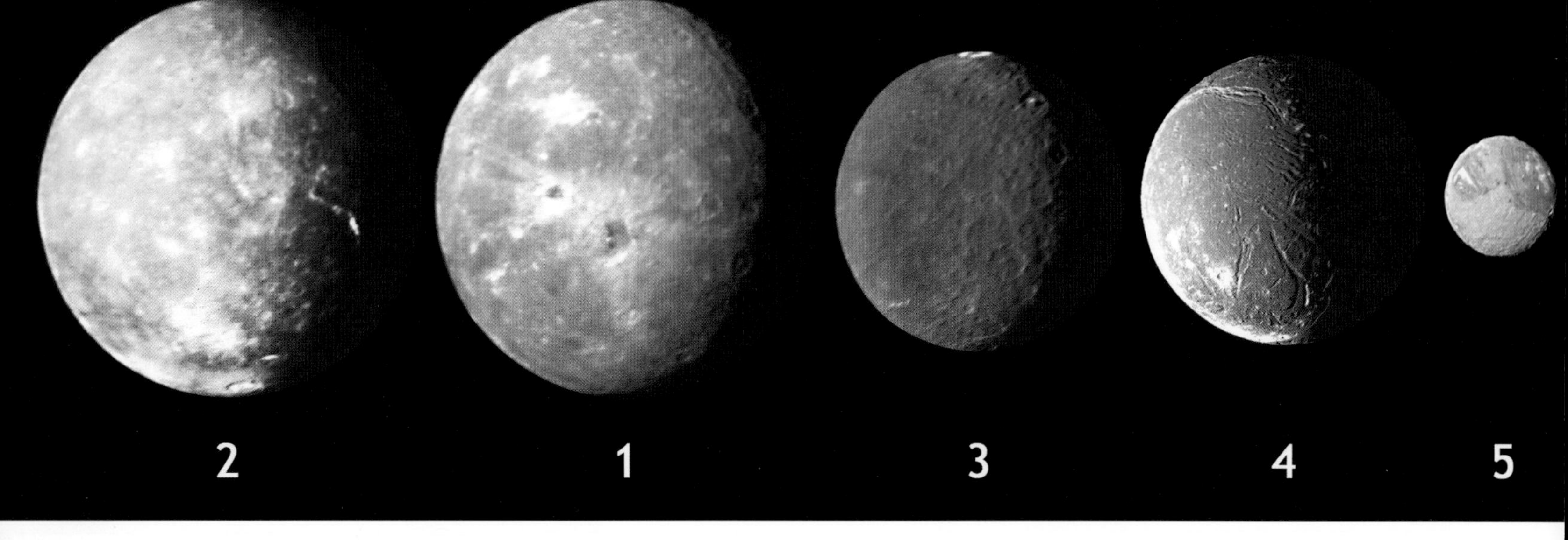

Image above: The five biggest moons of Uranus are shown here on the same relative scale. From left to right they are Titania, Oberon, Umbriel, Ariel and Miranda. The largest, Titania, is less than half the size of our own Moon. The numbers indicate the moons' positions relative to Uranus, with 1 (Oberon) being the furthest out. *Courtesy of NASA/JPL/Caltech.*

Aside from these classical satellites Uranus has at least 16 others. Eleven of them are located between Miranda and the rings, and they are regular – that is, they orbit in a similar plane and in the same direction in which the planet rotates. These moonlets truly are puny, between 13 kilometres and 77 kilometres across. They have dark surfaces, and compositions of ice and rock like their large cousins. Meanwhile, Uranus also has five irregular satellites. They orbit the planet at extreme inclinations, in different directions, and are very far from it – many millions of kilometres away. These moons, all of them again small, are all probably captured comets or icy planetesimals, and thus have different compositions from the regular satellites.

History of the Uranian System

As with Saturn and Jupiter, it is not easy to deduce Uranus' past because the planet has no solid surface, and thus no way of recording the events that may have shaped its evolution. Instead, astronomers rely on their models of the formation of the planets to deduce how Uranus formed and evolved.

The arrangement of Uranus' regular satellites – those which orbit in the planet's equatorial plane – hint that the Uranian system formed from a disc of material like the Solar Nebula in miniature, as did Jupiter and Saturn. However, if the Uranian system did form from a disc, how is it that it is now so tilted relative to the plane of the Solar Nebula defined by the orbits of the planets? Nobody has been able to answer this satisfactorily. But the most likely explanation is that Uranus – like Mercury, Venus and the Earth – suffered a shocking cataclysm, very early in its history. So the explanation goes, the icy planetesimal that would one day become Uranus was hit by another planetesimal in its neighbourhood. The crash knocked the Uranus planetesimal on its side, and the debris from the collision formed a disc around the tipped-up planetesimal. Later, when the planetesimal started to suck in gases from the Solar Nebula, these gases joined the sideways disc, and Uranus began to grow at the centre while its regular satellites coalesced in the disc's outer regions. Its remaining moons were captured at a later date. Uranus was unable to grow as massive as Jupiter because the Solar Nebula, as we have seen, was very sparse at this time. The planet has also lost some of its original hydrogen since it formed, because it has a weaker gravity than Jupiter.

The surfaces of the Uranian moons reveal that, even this far from the Sun, impacts were frequent – and often destructive. These impacts may also have helped to maintain the rings as they do with Jupiter, supplying the rings with dust-sized debris blasted off the moons' surfaces. The largest ring fragments, meanwhile, could be the remains of moons that got torn apart by Uranus' gravity. They might be cometary debris instead, but if this is the case then some process in the Uranian system must have darkened their surfaces.

Image below: From the rim of a crater on Titania, the full discs of Uranus and its three innermost large moons glow steadily overhead, something only visible once every 42 years. Although the planet appears about 13 times larger than our Moon does from Earth, it shines as bright as only two full Moons, so feeble is the sunlight.

Neptune – Last Giant Outpost

Perched almost on the edge of the realm of the planets, the serene blue Neptune is the last giant outpost. Another colossal world of slushy ice and gas, Neptune has many features in common with the green planet before it. But its hydrogen–helium atmosphere is one of constant turmoil, in contrast to that of the pallid and unchanging Uranus. Like the other giants, Neptune is graced by rings and satellites. Triton, its largest moon, is surprisingly active. It has a fresh, young surface. But something is very wrong with Neptune's comparatively small retinue of eight moons. If astronomers are correct, Triton is not a Neptune native – it was gravitationally captured. The process all but destroyed Neptune's original satellite system. And now, Triton aside, only haphazard shards of ice and rock remain in orbit about the blue giant.

Neptune Data

Mass: 1.02×10^{25} kg or 17.1 times Earth's

Equatorial diameter: 49 532 km or 3.9 times Earth's

Surface gravity: 1.77 gees

Axial tilt: 29.6°

Mean surface temperature: −220 Celsius

Rotation period: 16.12 hours

Orbital period: 164.8 years

Inclination of orbit to ecliptic: 1.8°

Orbital eccentricity: 0.009

Distance from the Sun: 29.79–30.32 AU

Sunlight strength: 0.0011 of Earth's

Satellites: > 8

Largest satellite: Triton, diameter 2706 km

Physical Overview

Image opposite: This impression shows Neptune's blue disc as it appeared to the Voyager 2 probe in 1989. Since then, however, the planet's appearance has changed. The conspicuous blue spot has vanished, replaced by others in different locations.

Taking almost 165 years to complete its orbit of the Sun, Neptune is extremely distant. From Earth there is not even the faintest chance of viewing it without binoculars or a telescope. And so, although it was discovered in the 1840s, virtually nothing was known about this fascinating blue gem for well over 100 years. Astronomers had to wait, twiddling their thumbs, until 1989 before they had any real data with which to work. This was the year in which the Voyager 2 probe made its last planetary encounter, before heading into the depths of interstellar space.

What Voyager 2 found was an apparent twin of Uranus, at least in terms of appearance. Neptune is slightly smaller than Uranus, by about 3 per cent, but it has a similar colour – again due to the presence of red-light-absorbing methane in its atmosphere. Even internally, Neptune is a lot like its green-faced partner. It has a rocky core – perhaps a bit larger than that of Uranus – surrounded by a vast ocean of icy slush, a thin shell of liquid hydrogen and other materials, and a hydrogen–helium atmosphere. But there is a surprise. While Uranus' atmosphere is bland and featureless, Neptune's is banded and stormy. This is perhaps odd. The blue planet is 1.5 times further from the Sun than its green neighbour, so its clouds ought to condense very low down, well out of sight. The reason for the difference is that Neptune, unlike its twin, has an unknown source of internal heat.

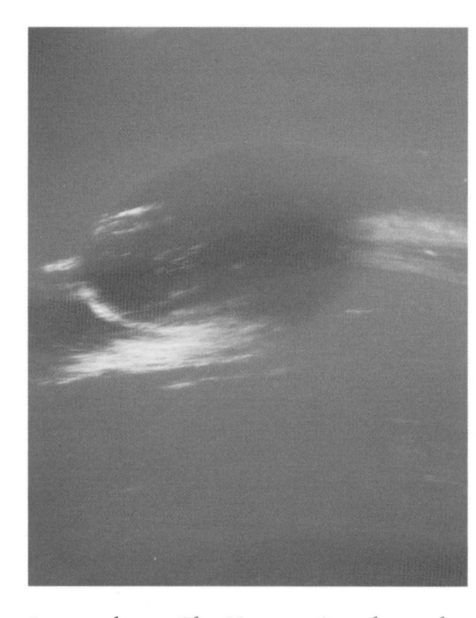

Image above: The Voyager 2 probe took this close-up of Neptune's Great Dark Spot from a distance of about 2.8 million kilometres. The spot seemed to be a storm system – like Jupiter's Great Red Spot – that rotates counter-clockwise, dragging wisps of cirrus cloud along for the ride. *Courtesy of NASA/JPL/Caltech.*

Image above right: The interior of Neptune is almost certainly much like Uranus'. There is a hot silicate core (grey), a thick ocean of dirty, slushy ice (white), a blanket of hydrogen, helium and other gases in liquid form (blue), all topped with a hydrogen–helium atmosphere. The Earth is shown on the same scale.

This keeps the planet warmer than it would otherwise be and drives the atmospheric activity so clearly evident on the planet's face. When Voyager 2 arrived at Neptune, the largest atmospheric feature on display was the Great Dark Spot. Scientists immediately likened it to a storm system, similar to the Great Red Spot on Jupiter. Scooting around this storm, meanwhile, were a number of wispy blue-white clouds of methane crystals. These floated some 50 kilometres above the bulk of the atmosphere, high enough to catch the sunlight. And yet, Neptune's great storm has since vanished. There was no sign of it in 1994 when the Hubble Space Telescope was swung towards the distant blue giant. Other cyclones, too, have come and gone. Evidently Neptune's weather patterns are not as long-lived as Jupiter's.

Rings

Not surprisingly Neptune has rings, just like the other three giants. And again, they are different from those we have looked at so far. As usual, Neptune's rings cannot compete with those of Saturn. They are very dim. Either they are made of rocky rather than icy fragments that reflect little light, or they are icy particles with a dark coating of organic molecules.

In total, Neptune has five rings. The most well defined, Adams and LeVerrier, are narrow, 50 kilometres and 110 kilometres in extent respectively. Still, these are generally wider than most of the rings of Uranus. The particles in these two rings are most likely metre-sized boulders. Two other rings are found between these two. One is again faint and narrow, while the other, though still faint, is fairly broad. Called Lassell, this ring stretches for about 4000 kilometres in radial extent. Lastly, the ring closest to the planet, Galle, is also fairly extensive, stretching for some 2500 kilometres. As well as the boulders present in some of these rings, all contain large amounts of dust, as found in the rings of Jupiter and Uranus. One thing that makes Neptune's rings stand out, though, are the so-called ring arcs. Most of the material in the rings is evenly distributed around their circumference, as it is in all other ring systems. But in three places, Neptune's outermost ring, Adams, is brighter than elsewhere. Apparently these arc-shaped segments contain more than their fair share of particles. At first this was a mystery. Why didn't these arcs spread out along their orbit and repopulate the rest of the ring? The answer, astronomers think, is that the arcs are caused by the gravitational pull of a small moon just inside the orbit of the Adams ring. This moon, called Galatea, is only about 160 kilometres across, with a puny gravity. But because it is so close to the Adams ring it is able to herd the ring fragments together and prevents them from spreading out along their orbit. Galatea is one of eight satellites around Neptune – fewer than the other giant planets.

Image above: This Voyager 2 close-up of two of Neptune's rings shows how in the outermost one, Adams, the fragments have strangely clumped together to form relatively dense arcs. The gravity of a small moon, called Galatea, is believed to be the cause. *Courtesy of NASA/JPL/Caltech.*

Triton and Other Satellites

Like Saturn, Neptune has only one large satellite. It is called Triton, and it is about 78 per cent the size of our own Moon. Its density indicates a roughly 50 : 50 mixture of rock and ice. The surface is dominated by frozen nitrogen, methane, carbon monoxide and carbon dioxide, with water ice underneath.

Triton is an exceptional world. Perhaps most surprising of all is that this frigid snowball is volcanically active – even though it possesses the coldest surface temperature ever measured in the Solar System, at −235 Celsius. But its volcanoes do not spout lava. Sunlight penetrates Triton's transparent icy surface and heats up underground nitrogen. As the gas boils and its pressure rises, it escapes through cracks in the surface ice and is squirted several kilometres above the moon's surface into its atmosphere. This is another surprise, that Triton should have an atmosphere. It is one of very few moons that do. Like Saturn's Titan, Triton holds onto its gas shroud – nitrogen with traces of methane – because the cold gases move too sluggishly to escape the feeble gravity. But this primitive sky is a far cry from the atmospheres of the terrestrial planets. Some 100 000 times thinner than the air we breathe, Triton's gaseous envelope is almost a vacuum. Most likely, Triton's atmosphere derives from the surface, and the reason has to do with the moon's orbit. Triton encircles Neptune backwards, in a path that is highly inclined to its planet's equatorial plane – it is the only major satellite with such an irregular orbit. This means that Triton's poles, like those of Uranus, endure long seasons when first one, and then the other – more than 80 years later – is turned towards the Sun. This sets up a

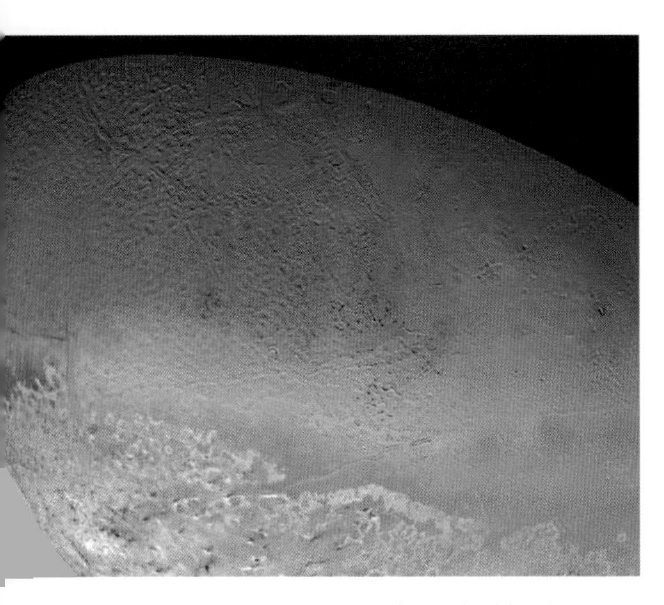

Image above: Far from the bland snowball astronomers were expecting, Triton's surface is remarkably young and still active. There are few impact craters, so the surface ices in all probability undergo regular freezing and melting. The black streaks near the bottom are caused by geysers or cryovolcanoes of nitrogen gas. *Courtesy of NASA/JPL/Caltech.*

Image opposite: A cryo-volcano on Triton spews a column of nitrogen gas and dirty organic ices high into the moon's thin atmosphere where the prevailing winds will blow it across the land. Elsewhere, similar events have left dark streaks that stretch across the surface.

cycle on Triton in which surface gases evaporate in the summer and help maintain the planet's atmosphere, then freeze out in the winter. This also explains some of Triton's surface features. There are few craters, probably because they are eradicated when liquid rises from the interior and later freezes.

While Triton is impressive, Neptune's other satellites are all little more than broken lumps of ice and rock. The largest of these worldlets is Proteus. It measures 400 kilometres by 440 kilometres, and so is not a perfect sphere. Meanwhile the smallest moons, closest to the planet, have sizes measured in just tens of kilometres. All but one of these moonlets lie well inside the orbit of Triton and, unlike that satellite, more or less in Neptune's equatorial plane. But one of the small moons chooses not to conform: Nereid. Some 340 kilometres across, Nereid has a highly inclined orbit, like Triton, and a very elongated one – its distance from Neptune varies by an astonishing 8 million kilometres, from 9.5 million to 1.3 million kilometres. Such an orbit is evidence, as is that of Triton, that something very dramatic must have happened long ago in the history of the Neptunian system.

History of the Neptunian System

So why is Neptune's satellite system in such disarray? Why is Triton's orbit so bizarre? The answer seems to be that Triton did not form around Neptune in a disc, as did the regular satellites of the other giant worlds. Neptune itself did grow from a disc of gas and ice just as Jupiter, Saturn and Uranus did. It is also possible that a system of regular satellites formed around Neptune as the planet was developing. But at some later time, Triton – a massive, icy planetesimal in a similar orbit to Neptune's – ventured too close to the blue giant and got captured into its retrograde and highly inclined orbit. During the event, Triton may well have collided with any regular satellites that had already existed. Either they were ejected from their orbits, or they were slingshotted into their parent world and destroyed. Now, aside from Triton itself, only shards remain. A few still orbit in Neptune's equatorial plane; perhaps these are the remains of its original satellites. Nereid, meanwhile, was almost ejected from the system – but not quite. Evidently it did not acquire enough speed to escape entirely, and instead settled into its very elongated elliptical meander.

As with the rings of Uranus, Neptune's are very likely recent. They probably did not form alongside their planet, for they would long ago have disappeared. Again, they could be fragments of comets or small moons that were torn apart by Neptune's gravity.

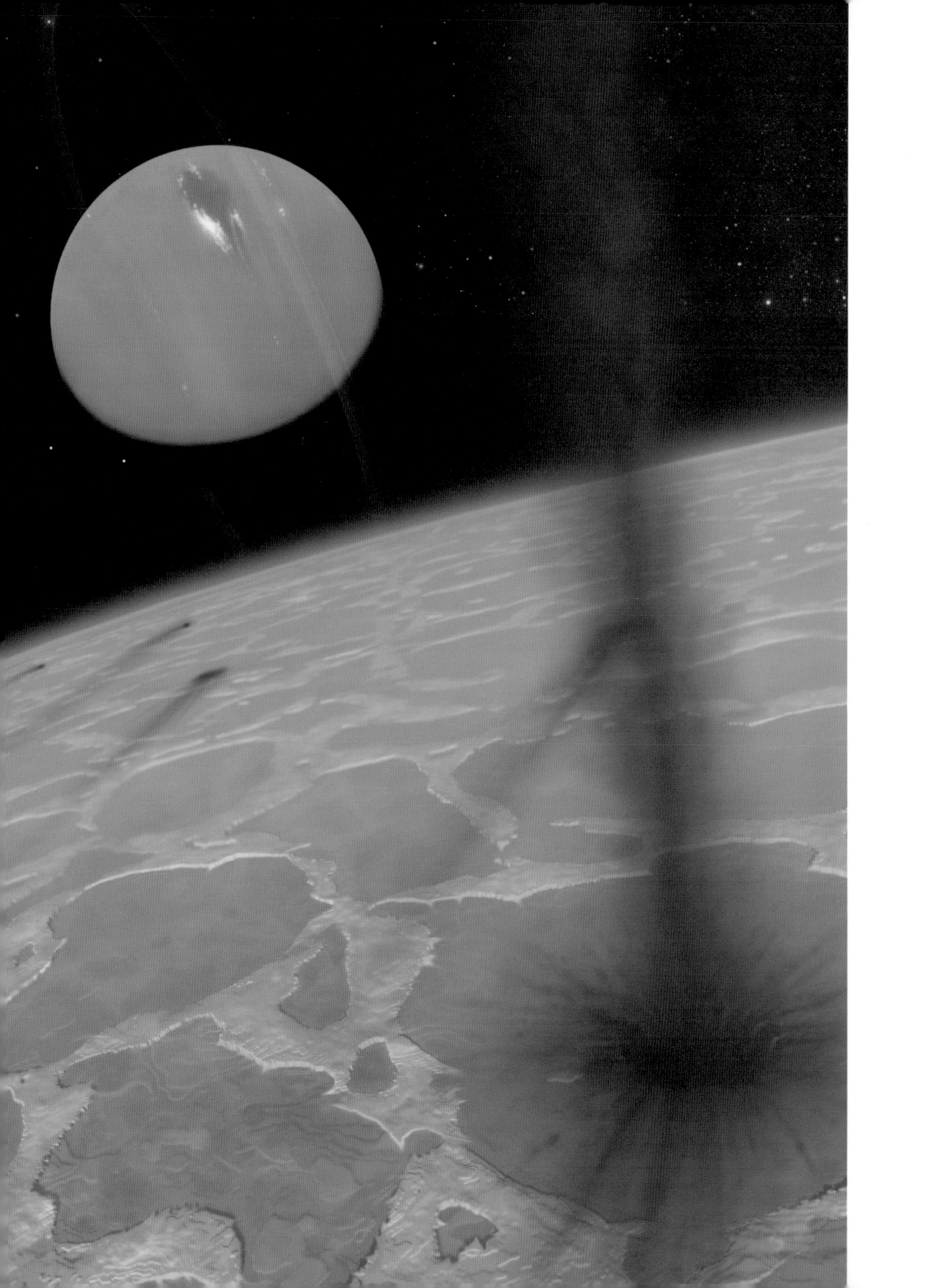

Pluto and Charon – Binary Planet

Pluto is the furthest known planet. At the most distant point in its orbit, even sunlight takes nearly seven hours to get there – and a car journey at 70 miles per hour would take well over a quarter of a million years to cover the same stretch! It is for this reason, along with the planet's very small size – 18 per cent that of the Earth – that astronomers still know very little indeed about this enigmatic worldlet. We do know at least that Pluto is primarily rocky with smaller quantities of ice. It also has a moon, called Charon. Fully half the size of Pluto itself, Charon is easily the largest satellite in comparison to its parent, and the pair has been called a binary planet. Pluto is an odd world – neither terrestrial nor giant. Indeed, some astronomers choose not to regard it as a planet at all on account of its diminutive size and its highly inclined and elliptical orbit compared with all the other planets. It seems more likely than Pluto is the largest of several icy worlds in the backwaters of the Solar System, out beyond Neptune.

Pluto Data

Mass: 1.32×10^{22} kg or 0.002 of Earth's

Diameter: 2300 km or 0.18 of Earth's

Surface gravity: 0.06 gee

Axial tilt: 119.6°

Mean surface temperature: about −230 Celsius

Rotation period: 6.39 days

Orbital period: 247.7 years

Inclination of orbit to ecliptic: 17.14°

Orbital eccentricity: 0.249

Distance from the Sun: 29.58–49.30 AU

Sunlight strength: 0.00041–0.0011 of Earth's

Satellites: 1

Charon Data

Mass: 1.6×10^{21} kg or 0.0002 of Earth's

Diameter: 1250 km or 0.098 of Earth's

Surface gravity: 0.03 gee

Mean surface temperature: about −230 Celsius

Rotation period: 6.39 days

Orbital period: 6.39 days

Inclination of orbit to Pluto equator: 0.0°

Orbital eccentricity: 0.01

Distance from Pluto: 19 636 km or 8.5 Pluto diameters

Pluto

Image opposite: Far from the Sun, Pluto (bottom right) and Charon revolve around each other like a giant dumbbell in space, invisibly tethered together by gravity. The surfaces of both worlds are very uncertain. But there is reason to suppose that Pluto could look somewhat like Neptune's moon Triton – complete with nitrogen geysers.

Pluto is easily the smallest planet. At 2300 kilometres across it is less than half the size of Mercury and only 70 per cent the size of our Moon. With a relatively high density – surprisingly so, in fact, considering Pluto's distance from the snow line – it must be mostly made of rock, with about 30 per cent ice. But no probe has ever been there, so very little is known about its surface. It is so distant that even the Hubble Telescope has great difficulty imaging the planet. Still, the best pictures show beyond reasonable doubt that Pluto's surface displays great contrast. The equator is dark with bright

Image above: Pluto's interior is still uncertain. But this world must have a fairly large rocky core because of its high density. This may be surrounded by a thick shell of water ice, and topped with scatterings of other frozen gases such as nitrogen, methane and carbon monoxide. Charon's structure is probably much the same but with a smaller relative core. The inset shows Pluto, Charon and the Earth on the same scale.

patches while the poles are light-coloured. One interpretation of this is that the poles are bright because they are covered in enormous polar caps of frozen methane. Meanwhile, other research has shown that ices such as nitrogen and carbon monoxide are also present on Pluto – very similar, in fact, to Neptune's moon Triton.

It is not only Pluto's size that makes it unique. Its orbit is tipped at an angle of 17 degrees to the ecliptic and it is more elongated than that of any other planet, even Mercury. This means that its distance from the Sun varies from 30 AU to 49 AU – a difference ten times the size of Earth's entire orbit. At its closest approach to the Sun, Pluto even crosses the orbit of Neptune. In fact from 1989 to 1999, Neptune was officially the most distant planet. But Pluto has now taken back its title and will hold onto it until well into the twenty-second century. Because of this bizarre meander, astronomers suspect that Pluto's appearance changes dramatically as it journeys around the Sun. At present, for example, Pluto is almost as close to the Sun as it gets – it is 'warm' enough for some of its ices to have evaporated to form a tenuous but quite extensive atmosphere of nitrogen and methane, again as on Triton. This atmosphere extends at least 600 kilometres above the surface, but its pressure at the surface is very low – roughly what you'd encounter 80 kilometres above the Earth. Gradually, though, as the planet heads away from the Sun over the next few decades it is quite possible that most if not all of its atmosphere will freeze out and lightly dust the surface in a blue-white snow of nitrogen and methane.

Image above: This is the clearest image ever obtained of Pluto and its moon Charon. It was acquired with the Hubble Space Telescope. Even with Hubble's powerful optics the pair are little more than featureless discs, such is their distance. *Courtesy Alan Stern, Marc Buie, NASA and ESA.*

Charon

One other thing that makes Pluto stand out is its companion, Charon. Over half the size of its parent and only 8.5 Pluto diameters away, Charon is even bigger in comparison to its planet than the Moon is to the Earth. Charon has a lower density than Pluto, so it must have a greater proportion of ice. Its surface properties are even more uncertain than Pluto's, but water ice seems abundant while methane is not.

Pluto and Charon take 6.39 days to orbit each other. Charon revolves around Pluto's equatorial plane. But, like Uranus, Pluto is tipped almost on its side. This means that the orbit of Charon is tipped more than 90 degrees relative to the Sun. Meanwhile, the orbital period of 6.39 days is also the amount of time it takes each world to spin on its axis. And so, just as the Moon keeps the same face to the Earth at all times, Pluto and Charon are forever locked in a similar gravitational embrace. From the surface of Pluto, Charon never appears to move in the sky. Instead, it just hangs there, forever presenting the same face but still going through a series of Moon-like phases. The same is true of Pluto seen from Charon. One way to visualise the pair is as a giant dumbbell with different sized masses on each end, endlessly tumbling in the vastness of space. If you happened to be on the hemisphere of Pluto turned away from Charon, or vice versa, you would never even know about your world's faithful companion unless you journeyed to the other side.

History of Pluto and Charon

So why is Pluto such an oddball? Why does it have a highly inclined orbit and axial tilt? Why is it so tiny, when the planets before it are all giants? The answers to these questions are all very speculative. But astronomers have come up with a few explanations that paint a very colourful past for Pluto and its companion.

At first, Pluto was believed to be an escaped satellite of Neptune – a sister of Triton. Certainly Pluto and Triton are enough alike to make this an interesting idea. They have comparable surface compositions, densities, atmospheres and radii. But, despite this, the current consensus is that Pluto and Neptune have never been anywhere near each other – although

Image above: Computer enhancement of images taken of Pluto by the Hubble Space Telescope have enabled astronomers to make these crude maps of the planet's surface. The Plutonian landscape seems to have a great deal of surface contrast, with a dark equator and bright polar caps of methane ice. *Courtesy Alan Stern, Marc Buie, NASA, JPL, Caltech and ESA.*

their orbits do cross, the planets never actually interact because they are always in different parts of the Solar System from one another. Nevertheless, some scientists still think that Triton and Pluto do share a common origin. If this is true, Triton once orbited the Sun independently just as Pluto does today. Interestingly, since the 1990s astronomers have found a few hundred icy bodies orbiting near Pluto and beyond in a region that has become known as the Kuiper belt. Like the asteroids, the Kuiper-belt objects are essentially leftovers from the planet-building process – as we shall see in the next and final section of Part 3. Thus Pluto and Charon – and Triton before it was captured – may just be the largest members of a whole family of icy worlds that never quite made the grade in the race to become planets, back at the dawn of the Solar System. In a sense, Pluto and Charon could be little more than large icy planetesimals, not a real planet and moon at all.

This is all very well, but it doesn't explain Pluto's weird orbit. To do that, we have to invoke some sort of cataclysm long ago in Pluto's deepest past. It should be obvious by now that the planets were frequent targets in the early Solar System's cosmic pool table. The Earth, Venus, Mercury and Uranus all show evidence of having been hit by something very, very big. We have seen how Mercury lost much of its mantle and become an iron world; how the Earth gained a satellite; that Venus was knocked upside down while Uranus ended up on its side. In Pluto's case, as with Uranus, perhaps a similar devastating collision with a neighbouring protoplanet knocked the world virtually on its side and also left it with its highly elongated and inclined orbit. Not only that but the event could also explain why Pluto has so much rock. It lost much of its icy mantle during its fatal encounter, just as Mercury lost its rock. Moreover, this scenario also offers an explanation for Charon's presence. The Pluto collision would no doubt have left a lot of debris, and Charon could be the product of the accretion of that debris in orbit around Pluto. Strange to think that, so far from the Sun, Charon might be the outcome of the same mechanism that produced our own Moon.

Comets – Dirty Snowballs

Just beyond Pluto and Charon is the first of two gigantic reservoirs – a donut-shaped region of space known as the Kuiper belt. As with the asteroid belt, the contents of the Kuiper belt hail from the earliest stirrings of the Solar System. But the planetesimals found this far from the Sun are not stony or metallic like the asteroids. Instead, the Kuiper belt is home to thousands of balls of dirty ice – cometary nuclei, frozen solid. Beyond the Kuiper belt, flattened nearer the Sun but spherical further out, is an even larger reservoir: the so-called Oort cloud. If calculations are correct, the Oort cloud comets have a population measured in trillions. Nobody is certain how far this phantom cloud extends. But it is quite likely that the Solar System's most distant comets orbit the Sun more than fifty thousand times further out than our humble Earth. This is the true edge of the Solar System.

Kuiper Belt Data

Innermost edge: 30 AU

Outermost edge: 1000 AU?

Orbital period at innermost edge: 165 years

Orbital period at 1000 AU: 32 000 years

Known population: $>$ 380

Estimated total population: 200 million?

Estimated total mass: 4.8×10^{23} kg 0.08 Earth's

Composition of bodies: ices with some rock

Largest member: 2001 KX76, diameter 1300 km

Oort Cloud Data

Innermost edge: blends into outer Kuiper belt

Outermost edge: tens of thousands of AU

Orbital period at outermost edge: millions of years

Estimated total population: 6 trillion

Estimated total mass: 40 times Earth's

Composition of bodies: ices with some rock

Comets

Image opposite: Though comets themselves are only kilometres across, the gas and dust freed from their icy surfaces can stretch very long distances, blown by the solar wind and by radiation pressure. Here, a bright comet passes behind the Earth as it heads towards the Sun, its tail – longer than the radius of Earth's orbit – dwarfing the puny planet and its satellite.

Like the asteroids, comets are ancient. They are the incomplete fragments of the accretion process that built our Solar System billions of years ago. But, despite their common origin, the asteroids and comets are very different classes of objects.

While asteroids are stony and metallic, comets are composed primarily of ices and rock – they have been called 'dirty snowballs'. This is because they were formed much further from the Sun, way beyond the snow line. Most of the ice is frozen water, but comets also contain ices of carbon dioxide, carbon monoxide and methane. However, comets have a very low density – about one-quarter that of ordinary water ice – and because of this they must be extremely porous. Either their ices are loosely packed, or they have large numbers of voids within them, pockets filled with gas under pressure. This makes them very fragile, which was why the comet Shoemaker-Levy 9 was broken into pieces so easily by Jupiter's gravity. Another difference between these dirty snowballs and the asteroids is in

Image opposite: An impression of what it might be like to peer past the storm of gas, dust and ice particles and into the heart of a comet. Jets of gas are seen escaping from the sunward side, providing the source for the comet's tail and coma. For comparison, the inset shows a photo of the nucleus of Comet Halley taken by the European Giotto probe.

their visual appearance, and that has a lot to do with their orbits. Comets have very elliptical paths that take them alternately far from the Sun and then close in. When furthest from the Sun, at aphelion, comets are frozen solid and very hard to see. Their dirty, puny surfaces – most are just kilometres across – reflect very little light. But, as they near the Sun, their ices start to sublimate – and a miraculous change occurs.

Freed from their frozen prison, the gas and dust surround the comet – now more properly called the comet's nucleus – in an expansive but very diffuse shroud known as a coma. Jets of gas escaping from voids in the warming nucleus also feed the coma. At perihelion, the closest approach to the Sun, a typical coma may measure one million kilometres across – larger even than our star but of course much thinner. Vastly bigger still is the comet's tail. Electromagnetic forces exerted by the solar wind rip charged particles out of the coma and blow them away from the Sun to form the comet's ion tail. The dust, meanwhile, is similarly pushed outwards by the pressure of the Sun's own radiation. Thus, cometary tails always point away from the Sun. The dust reflects sunlight and appears yellow, whereas the ions emit primarily blue light as they recombine. Together, the dust and ion tails can stretch for well over a hundred million miles – comparable to the radii of the orbits of the inner planets. And then, as the comet moves away from the Sun and into the depths of interplanetary space, it slowly begins to freeze once more, and gradually fades from view – until next time.

The Oort Cloud

Because of this constant freezing and thawing, comets lose a great deal of material with every orbit. Thus even the largest comets must have finite lifetimes, beyond which they are either evaporated away completely or reduced to inert balls of rock. Estimates show that some might last for only 10 000 years or so. How is it, then, that the comets are still around? If they were created in the early stages of the Solar System but only last for tens of thousands of years, why were they not all extinguished billions of years ago?

A clue to answering this question can be obtained by looking more closely at cometary orbits. In general, comets come in two classes. There is the short-period variety, which orbit the Sun in less than 200 years; and there are the long-period comets, which take longer to complete their journey. These definitions are not quite arbitrary. While all cometary obits are generally elliptical, those of the short-period type are generally well constrained to the ecliptic. The long-period comets, by contrast, can have orbits that are greatly inclined to the ecliptic – any inclination at all, in fact. The path of Hale-Bopp, for example – the bright comet seen in the first half of 1997 – intersected the Solar System at right angles. Because these long-period comets can penetrate the realm of the planets from any direction, a vast swarm of them must surround our Solar System in a shell or

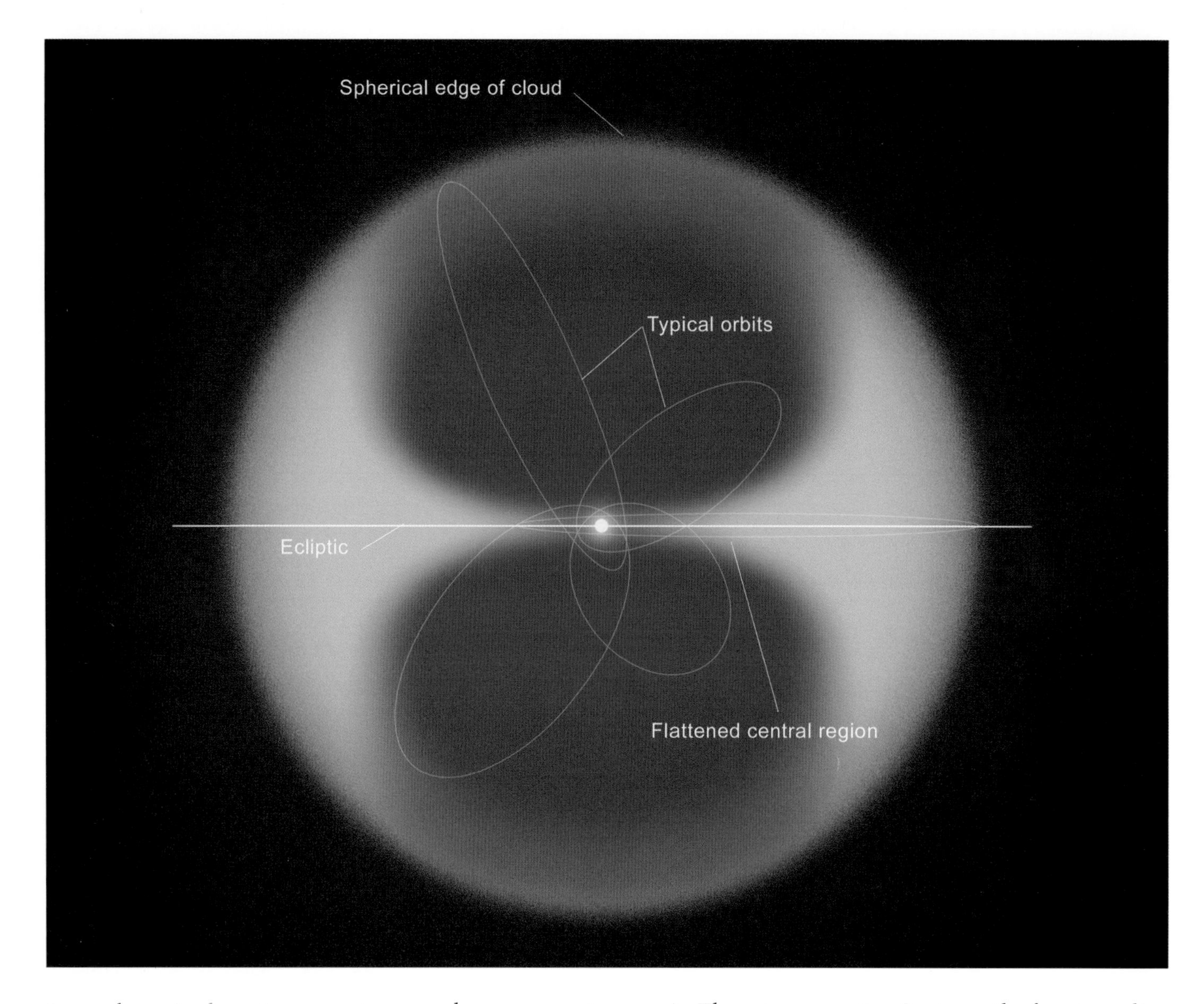

Image above: A schematic representation of the Oort cloud with the main features exaggerated for clarity. The outer boundary is spherical and diffuse. Comets entering the inner Solar System from here can do so at a range of angles, as shown. Nearer to the Sun the comets are found increasingly closer to the ecliptic. The orbits of the planets are too small to be seen on the scale of this diagram, and even the Kuiper belt (not shown) is only a few millimetres across at most.

sphere – a comet reservoir. The astronomer Jan Oort was the first to realise this, in 1950, and this hypothetical storage tank has become known as the Oort cloud in his memory. Oort originally envisioned a fog of countless icy fragments surrounding the Sun at distances of 100 000–300 000 AU from the Sun, within reach of the nearest stars. From time to time, some of them would experience gravitational perturbations, at the mercy of passing stars. These comets would then be flung into more distant orbits – or towards the Sun where they would be captured into short-period, elliptical orbits. This helped explain how comets could last so long – they were constantly being injected into the Solar System to replace those worn out by millennia of repeated thawing.

Modern astronomers have accepted Oort's idea. But there have been some refinements. In particular, the Oort cloud is now seen as somewhat smaller than originally thought – perhaps only 50 000 AU (0.8 light-years) in radius, but still enormous in comparison to the orbit of Pluto. In total it is estimated that there could be 6 trillion comets there, with a combined mass nearly 40 times that of the Earth.

The Kuiper Belt

Still, the Oort cloud is not the complete picture. In the late 1980s, a trio of astronomers at the University of Toronto showed that most short-period comets could not possibly have been injected into their orbits from such far off distances. Recall that these comets have orbits near the ecliptic. Perhaps, instead, they were coming from a smaller, flattened disc surrounding the Sun beyond the orbit of Neptune. Interestingly, the existence of a trans-Neptunian band of icy planetesimals had first been postulated at around the same time the Oort cloud was – by Kenneth Edgeworth in 1949, and Gerard Kuiper two years later. For obvious reasons, this has become known as the Kuiper or the Edgeworth–Kuiper belt.

Despite the indications in the 1980s that some comets hailed from the then hypothetical Kuiper belt, astronomers didn't actually discover the first trans-Neptunian object to support the theory until 1992. Called 1992 QB1, it is a relatively large, dark body about 250 kilometres across, roughly 41 AU from the Sun. Nowadays, the total number of worldlets known to orbit in the icy wastes beyond Neptune has crept above 400, and is growing all the time. The Kuiper belt is as real as the asteroid belt.

Studies of the orbits of the Kuiper-belt objects (KBOs or TNOs for trans-Neptunian objects) have revealed that they fall into three distinct classes. By far the majority of the KBOs – the so-called classical TNOs – have near-circular orbits and are found more than 40 AU from the Sun. Their orbits are generally close to the ecliptic plane. Others, about 30 per cent in all, have elliptical orbits that are instead mildly inclined to the ecliptic, at average distances from the Sun of about 39 AU. These particular members share another characteristic. They orbit the Sun roughly twice in the time it takes Neptune to go around three times. These orbits are thus said to be in a 2 : 3 resonance with Neptune's. (Oddly enough, Pluto has just such an orbit – in fact those KBOs with similar orbits are often called 'plutinos'. If Pluto were discovered today it would almost certainly be regarded not as a planet but as a particularly large KBO.) The remaining KBOs, a few dozen, all have highly inclined orbits, but they are in addition extremely elliptical – some have orbital periods measured in tens of thousands of years.

It is well established now that the Kuiper belt is the source of many of the short-period comets; the long-period ones all come from the Oort cloud. However, the exact extent of the Kuiper belt and its population is very uncertain. It was originally thought that it might extend for 100–1000 AU, interfacing with the innermost regions of the Oort cloud. And yet, despite the fact that new KBOs are found all the time, so far none has been found with an average distance of more than 55 AU from the Sun. This could be because we just do not have sensitive enough equipment – without exception the Kuiper-belt objects are very dark and exceedingly difficult to see even with large telescopes. This is why it took so long to find them despite the fact that some of them are very large. (The biggest

Image above: Comets often come dangerously close to the Sun – and sometimes even fall into it. This image, captured by the LASCO instrument on board ESA's Solar and Heliospheric Observatory (SOHO), shows one such comet, dubbed a sungrazer (left). The Sun is in the middle (white circle) shielded by the larger orange disc. *Courtesy SOHO/LASCO Consortium.*

KBO to date, called 2001 KX76, is an incredible 1300 kilometres across. This is much larger than the biggest asteroid, Ceres, and even exceeds the diameter of Pluto's moon Charon.) Certainly the number of Kuiper-belt comets is far less than those in the Oort cloud. It may be measured in the hundreds of millions, and their combined mass is probably only a small percentage of the mass of the Earth. This is enough to make a planet intermediate in size between Mercury and Mars – still more than could be salvaged from the asteroid belt.

History of the Comet Reservoirs

Like the asteroids, the comets in both reservoirs are throwbacks to the dawn of the Solar System, the remains of, primarily, the outermost environs of the Solar Nebula. Far beyond the snow line, the Solar Nebula gathered its ices together to form the outermost giant planets Uranus and Neptune. Once these appeared, they started to behave like bullies in a school playground, their gravities stirring up the nearest planetesimals. The largest icy planetesimals were sufficiently massive to resist being flung far from the realm of the planets. Instead, they remained close to where they formed. They included Pluto, Charon, Triton and some of the largest modern-day KBOs. Were it not for Neptune's ever-present gravitational tugs, these bodies might have coalesced to form a single, larger planet – but by no means a giant.

Meanwhile, the smaller icy planetesimals suffered a different fate. They were flung far from the Solar System where the great majority settled in a flattened disc that surrounded it at a great distance – tens of thousands of astronomical units. Some may even have been sent out of the plane of the Solar System; but it takes a great deal of energy to do this, and neither Neptune nor Uranus had sufficient mass to do it very often. Nevertheless, the comet distribution is now very different. And for that, nearby stars are to blame. Since the birth of the Solar System, the Sun has weaved an intricate path for itself – and its cargo of planets and comets – through interstellar space. On occasion it has come very close to its stellar neighbours – perhaps even as close as a few hundred astronomical units – and on the whole the Sun will pass within 10 000 AU of a nearby star every few tens of millions of years. That adds up to hundreds of thousands of encounters in the Sun's

Image right: In 1994, astronomers watched, astonished, as fragments of a comet broken up by Jupiter's gravity plunged into that planet to meet a fiery doom. This image taken by the Hubble Space Telescope shows the comet, Shoemaker-Levy 9, as a train of 21 icy fragments two months before the fateful encounter. In total, the fragments in this image spanned a distance three times that from the Earth to the Moon. *Courtesy H. A. Weaver, T. E. Smith (STScI), and NASA.*

4500-million-year history. Gravitational perturbations exerted by the passing stars during these fly-bys have scattered the outermost edges of the original comet disc so substantially that they now form a sphere rather than a disc – the well-known Oort cloud. Meanwhile, those comets closest to the Sun, having suffered fewer perturbations, still define their original donut-shaped volume. This is the inner core of the Oort cloud, which blends into the Kuiper belt in its centre.

Not all of the objects in the comet clouds originated near Neptune, however. Jupiter, having a much greater mass than Neptune, no doubt provided quite a bit of muscle in the early days of the Solar System when the comet clouds were forming. In fact, Jupiter is so massive that most of its casualties escaped the Sun completely. But despite this, the richness of the protoplanetary disc near Jupiter meant that a sizeable fraction of its planetesimal neighbours did end up in the Oort cloud and Kuiper belt. These planetesimals, originating nearer the Sun, contained larger quantities of rock than those that hailed from the vicinity of Neptune. And so, quite possibly, not all of the objects in the comet reservoirs are pure ice – there may be a substantial asteroid population there too.

Image above: Comets are basically lumps of dirty ice (brightest blobs) loosely packed together and covered by a very dark crust. The low densities of cometary nuclei indicate that they must be porous, perhaps with large empty voids (black) inside. The largest comets of all may in addition have small rocky cores (not shown).

Part 4
End of an Era

'The bright day is done,
And we are for the dark.'

William Shakespeare, *Anthony and Cleopatra*, V.2

We have seen how the Solar System came to be, and how it has changed in the billions of years since it was born. Now it is time to take a different journey – a journey into the future of the Solar System. This is the subject of Part 4.

We think of the Sun as all-powerful and everlasting. Indeed, on a human timescale it is. Deep within its fiery interior, though, the numbers speak for themselves. On the main sequence the Sun converts a phenomenal 600 million tonnes – the mass of a small mountain – of hydrogen into helium every second, just to keep itself balanced against gravity. At the moment, there is no need for us to worry about this alarming appetite. For the Sun has enough hydrogen to keep its nuclear fires stoked for a good few billion years into the future – long after mankind has vanished. But the day will come when the Sun's fuel heap will run dry, and its useable hydrogen has been totally consumed. When that happens, the Sun will start to die – and with it, the rest of the Solar System.

All astronomers agree on the broad details of what will happen to the Sun. As its hydrogen runs out, it will swell, slowly at first, to become a subgiant – then much more rapidly as it ascends the red giant phase, becomes a planetary nebula, and ends up as a tiny ember called a white dwarf. We shall see these details for ourselves within the next few pages. These are as certain as things can be in astronomy. But what is not quite so established – as was also the case with the building of the planets – is the timeframe involved. Some researchers place the main-sequence lifetime of the Sun at 9 or 10 billion years, while for others a number closer to 13 billion is more acceptable. The dates that appear on the following pages, then, are not necessarily definitive, but are intended to give a flavour of things to come. The model I adopt has a main-sequence lifetime of 10.9 billion years. With the current age of the Sun, 4600 million years, this means that our star has about 6300 million years to go before it experiences any major structural changes. Already, then, the Sun is almost halfway through its main-sequence life – to the beginning of the end of an era.

Present-day–10 900 million years
Main Sequence

Image opposite: We are 1 billion years in the future, and the Sun is 10 per cent brighter than the star we once knew. Dry river beds, like the one shown here, are the norm rather than the exception, and the planet Earth has become a hot, humid graveyard for trees and large animals.

The Sun is already dying, in a sense. All the while our star burns hydrogen on the main sequence, its core becomes gradually more depleted in that element, and a helium 'ash' is left in its place. As the core adjusts itself to this steadily changing composition, the star's diameter and brightness both slowly increase. When the Sun took its first steps onto the main sequence it was only 90 per cent of its current radius and 60 or 70 per cent of its present luminosity. It is quite a bit warmer and larger now than it used to be. And that trend is not going to change.

The next billion years will see a hike in the Sun's luminosity by about 10 per cent. That may not sound like a cause for concern, but for the innermost planets the change will be overwhelming. And, for the Earth in particular, this slight escalation in luminosity will signal the beginning of the end of billions of years of evolution. With that much extra energy flowing away from the Sun, our planet's polar caps will start to melt and its oceans will begin to warm up. Slowly, they'll evaporate into the atmosphere. Too much water vapour, like carbon dioxide, has a serious effect on our planet's climate. The Sun's energy heats the surface, but the heat is partially trapped. Infrared radiation cannot travel through water vapour or carbon dioxide, because they absorb it. And so the planet steadily warms up. Today, Earth is about 32 Celsius warmer than it would be without its atmospheric blanket. But, a billion years from now, when the oceans and polar caps have begun to vanish, Earth's global temperature will rise even higher. The planet will grow hotter and more humid – a situation known as a moist greenhouse effect. The results will be slow at first. But continued heating will render the Earth's surface uninhabitable, or at least hostile to life, by the time the Sun reaches its six-billionth birthday. With the surface of the Earth tens of degrees hotter on average than at present, those oceans and seas that remain will be the only safe havens for the lifeforms that survive – but only for the next 2 to 3 billion years.

After that, about 3500 million years from now, the Sun will be much brighter still – 40 per cent more luminous than now. The oceans start to disappear on a gigantic scale, totally lost to outer space. Carbon dioxide, without the oceans to dissolve in, accumulates in the atmosphere as it did on Venus billions of years ago – and Earth suffers the same devastating runaway greenhouse effect that Venus has endured for countless millennia. The planet Earth becomes bone dry, with a surface temperature of hundreds of degrees, and lifeless aside from a few hardy forms of bacteria.

This, though, is still long before the Sun nears the end of its fuel. That time will come a further 2500 million years on into the future, more than 6 billion years from now. The Sun will then be about 10 900 million years old. Its long period of stability will finally be at an end.

10 900–11 600 million years
Subgiant Phase

Image below: Seen from a lifeless, desert Earth now devoid of an atmosphere, 7 billion years from now, the alien Sun appears larger, brighter and redder than the star we know. This is a star called a subgiant.

Image opposite: The subgiant Sun (right) compared with its modern-day cousin. The subgiant is substantially larger than the main-sequence Sun and somewhat cooler and therefore redder.

After a main-sequence lifetime of almost 11 billion years, the Sun finally reaches the end of the hydrogen in its core. Its nuclear fires, once seemingly endless, now cease. Helium does not burn at the relatively low temperatures needed to fuse hydrogen, because helium nuclei have double the electric charge of hydrogen nuclei. They have to move much faster to stick together, and this in turn requires a far greater temperature. Thus, without an internal power source to hold the star up against gravity, the delicate balance of hydrostatic equilibrium in the helium-choked core vanishes – and the core starts to shrink as it succumbs to its own gravity. This marks the beginning of the end for the ageing Sun.

The helium core itself is completely inert. But its contraction brings with it a new, brief, lease of life for the star. For a fresh supply of unused hydrogen – that which had originally resided on the outskirts of the core where it had been too cool to fuse – is now dragged inwards with the core's contraction. As the new fuel arrives on the edges of the core, its gravitational potential energy is converted into heat. The centre of the star becomes even hotter than it was before. Now this virginal hydrogen, at last hot enough to

undergo thermonuclear reactions, begins to do just that. It burns in a thin shell surrounding the lifeless core of helium. Energy anew streams outward from the centre of the star and pours into the outer layers of the Sun, the pressure of the radiation pushing them away slightly. And so, while the core slowly shrinks, the Sun as a whole expands. The outermost parts, now further from the heat-generating core, thus begin to cool down slightly as they creep outwards. This heralds the start of a relatively rapid transformation, lasting around 700 million years. As these changes set in, the Sun is already about 1.6 times larger than at present, somewhat redder, and more than twice as luminous – brighter because of the increased surface area. However, with hydrogen burning only on the outskirts of the core, the core itself remains unsupported against gravity and continues to contract – and the Sun grows steadily larger and redder all the time.

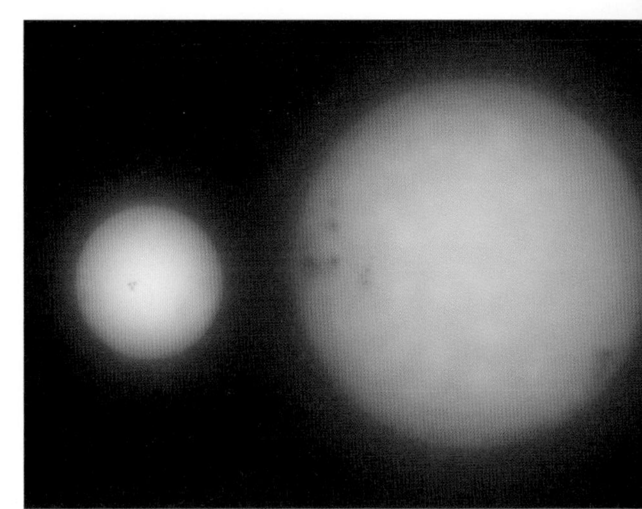

When this period ends, more than 3 billion years after the end of life on Earth, the Sun is around two to three times its current diameter and about 800 degrees cooler. The star we know has vanished forever. It has become a new type of star – one known as a subgiant.

11 600–12 233 million years
Red Giant Phase

Image below: The red giant Sun is close to its maximum size and reaches 70 per cent of the way to scorched, barren Earth. Seen from there the swollen star appears impossibly huge and bright, and spans 90 degrees of sky.

Image opposite: When the Sun swells to become a red giant it completely swallows Mercury. Only Venus and Earth escape a fiery death, having moved out to safer distances (indicated) as a result of the Sun's mass loss.

The subgiant phase is dramatic. But it is nothing in comparison to the metamorphosis that the Sun undergoes next. All the while the subgiant grows, its surface layers cool. This makes them more opaque to the radiation streaming through them – and eventually they become so opaque that the radiation generated in the core cannot easily escape into space. The Sun will face this crisis at an age of 11.6 billion years, when its surface has cooled to about 4900 Celsius. The energy produced deep in its interior is trapped. It builds up, and the resultant pressure drives the bubbling, convecting surface layers away faster and much further than ever before. In just 600 million years the Sun swells to 160 times its current diameter. And its surface, having cooled down even further to about 3100 Celsius, is now distinctly red. But red and cool do not mean faint. The enormous increase in surface area makes the monster Sun exceptionally luminous. More than 2000 times brighter than its main-sequence luminosity, such a star is known as a red giant.

The fate of the inner planets at this point is very drastic. At 160 times its present size, the red giant Sun extends out to about 0.7 AU. Mercury's destiny is clear. As the Sun grows ever larger, it completely engulfs the tiny planet. The Solar System's planet count reduces from nine to eight. Venus' future, though, is more uncertain. For the red giant phase is accompanied

by a tremendous stellar wind, far more powerful than the feeble wind the Sun has today. This stream of charged particles blusters away from the red giant with such ferocity that within a few tens of millions of years of reaching the red giant phase the Sun will shed about 30 per cent of its mass. With a smaller gravitational well at the centre of the Solar System, all of the planets move outwards slightly in their orbits. Mercury has gone. But Venus moves to about 1 AU, the current location of the Earth, while the Earth itself occupies a new orbit somewhere near where Mars is today, and so on. These orbital shifts mean that Venus might at this point escape the fate of Mercury. Earth almost certainly will. But even at 1 AU, Venus could well be orbiting inside the outer atmosphere of the Sun. If that is the case, atmospheric drag will inexorably decay the planet's orbit – just as the Earth's atmosphere de-orbits artificial satellites today – and pull it down to embrace the same fiery doom that its neighbour suffered earlier.

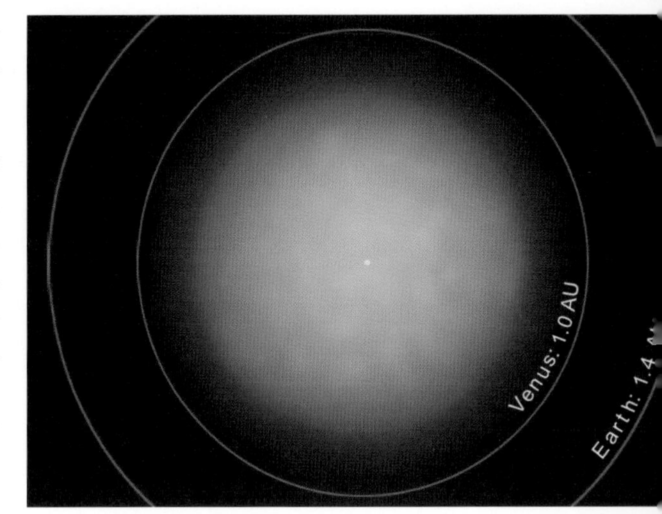

Just as during the subgiant phase, hydrogen consumption in a red giant takes place not in the centre of the core, but around the edges. So all the while the Sun's atmosphere is expanding, its unsupported core grows gradually smaller – and hotter. This continues until, at an age of about 12 233 million years, the core temperature reaches an all-time-high – and the stage is set for another major change.

12 233–12 365 million years
Helium Burning and Second Red Giant Phase

Image above: Most stars are so far away that they appear as specks even with the most powerful telescopes. But Betelgeuse shown here, a red supergiant in Orion, is an exception. This Hubble image shows the star's actual disc. *Courtesy Andrea Dupree, Ronald Gilliland (STScI), NASA and ESA.*

Image opposite: Seen from the surface of Jupiter's moon Europa, the red giant Sun appears almost as large in the sky as the giant planet itself. The ice that once covered the moon has now melted, and life perhaps thrives in the ocean that covers the surface.

Eventually the core of the red giant Sun attains a temperature of 100 million degrees. This, at last, is hot enough to trigger more nuclear reactions. But this time the reactions feed on helium, not hydrogen. At this temperature, the helium nuclei within the core move so quickly that they overcome their electrostatic repulsion and combine to produce a new ash – carbon and oxygen. Within seconds, the entire core is alive once more, vigorously producing energy. This ignition of core helium is so violent and sudden that it is known as the helium flash. Suddenly the core's gravitational collapse is halted, and the Sun's expansion actually reverses. Within just one million years the entire star shrinks to about ten times its main-sequence diameter – down from its red-giant diameter of about 16 times that. With helium fusion taking place in the core and hydrogen burning continuing in a shell on the outskirts, hydrostatic equilibrium holds once more. The new slimline Sun enjoys a second taste of stability – for the first time in 1300 million years. But helium reactions happen at a much faster rate than hydrogen reactions, because of the higher temperature. The helium in the core is thus exhausted very quickly. In less than 110 million years, the Sun must once more face a crisis.

This time, though, the crisis will be fatal.

With no more fuel to burn, the core starts to fall inwards yet again. Just as happened when the original core ran out of hydrogen, the contraction warms the core and brings more reactionable fuel – this time, helium – into a hotter environment. Once more, helium starts to burn. These new reactions take place in a shell between the inert carbon/oxygen core and the shell of hydrogen reactions. Again, as was the case with the initiation of hydrogen-shell burning, the fresh energy produced in the helium shell floods into the opaque outer layers and the ill-fated star begins to expand towards a second red giant phase. But whereas the previous expansion to this state took 600 million years, this time it takes a mere 20 million years – and the result is even more dramatic.

At its maximum extent, the Sun will stretch to 180 times its main-sequence diameter – much larger compared with the Sun we know than the present Sun is compared with the Earth. It will be 3000 times brighter than at present. Then once again its wind picks up, and the red giant loses even more mass. Slowly, the planets edge outwards even more, with the Earth and maybe Venus, again, just escaping a fiery doom in the atmosphere of the turbulent red giant. Meanwhile, the ice moons that surround the more distant planets by this stage have warmed up to the extent that their ices have melted. Europa's icy shell might become an ocean of liquid water. Together with Saturn's moon Titan, it could provide a habitat for new life in the Solar System. But what life may have evolved by this point will never attain the rich complexity of life as we know it – for now, time truly has run out in the Solar System.

12 365 million years
Planetary Nebula Phase

Image opposite: More than 12 billion years after it formed, the Sun discards its outer layers and surrounds itself in a colourful shroud of nebulosity known as a planetary nebula. The core of the original star can still be seen at the centre of the shell.

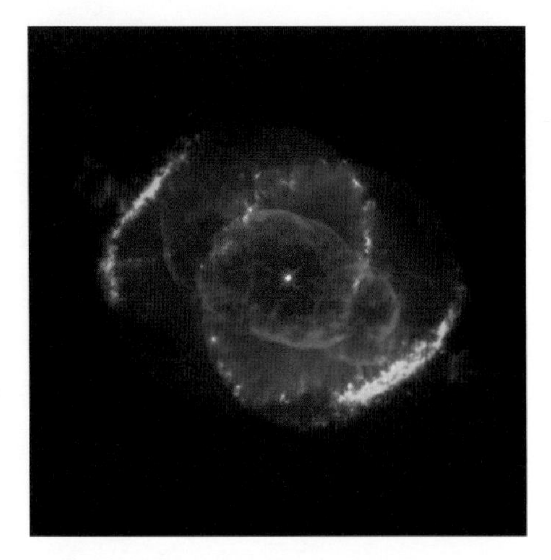

Image above: Not all planetary nebulae are simple shells. Some, as shown in this image of the Cat's Eye Nebula taken with the Hubble Space Telescope, are much more complex. The nature of this complexity is an active area of current research. *Courtesy J. P. Harrington, K. J. Borkowski and NASA.*

The Sun now has nothing left but shell burning to sustain it. But there is a problem. Helium burning is inherently unstable, because the rate at which it proceeds is exceedingly sensitive. A temperature rise of just 10 per cent means a hike in the reaction rate by a factor of 45! All it takes is a slight positive fluctuation in core temperature, and the result is a powerful pulse of energy that surges out through the already bloated Sun and puffs it up like a smoke ring – to even greater dimensions.

Simulations show that the Sun will swell beyond its main-sequence radius by a factor of more than 210 during such a pulse. At this truly phenomenal size the Sun encompasses the entire volume of space that at present exists between it and the third planet, Earth. It is 300 million kilometres across, and reaches a peak luminosity some 5200 times that of the star we know today. These circumstances exist only briefly, for maybe a few tens of thousands of years. For, when the core temperature suffers a negative fluctuation and starts to fall, the energy pulse dies away and the star shrinks again. According to astronomers' calculations, the red giant Sun is expected to suffer up to four of these thermal pulsations, one every 100 000 years. It is these that will finally lead to the demise of this truly ancient star. The reason is that the outermost layers of a red giant are very weakly bound. When an energy pulse surges through them, its momentum carries away the outermost material so quickly that the star's gravity cannot suck it back when the pulse is over. Moreover, because of the diminishing overall mass and therefore gravity of the star, each pulse removes even more material than the one before it. And so in this way, the star gradually and gently blows itself apart over the course of several hundred thousand years. Now the dead star is poised for perhaps its most spectacular performance.

After the fourth and final pulse, the Sun is left with only about half of its original mass. With the red giant's entire tenuous envelope now completely removed, only the central core remains, tiny, hot, naked. Waves of intense ultraviolet radiation stream outwards from the hot, exposed core and plough into the rapidly expanding layers of the Sun to ionise the gases within them. They respond, as their component atoms recombine, with a dazzling display of colour – the hallmark of a planetary nebula.

But sadly, although perhaps the most beautiful aspect of the Sun's entire life, the planetary nebula is also one of its most fleeting phases. Gradually, the discarded shell of gas continues to expand away from the central, ionising core, all the time growing dimmer and thinner. At its maximum extent the nebula might stretch for a few tenths of a light-year. But, by this time, 10 000 years after it forms, the nebula is too far from the hot core for effective ionisation, and thus too faint to remain visible. And, after the dispersal of the planetary nebula, only the brilliant core remains.

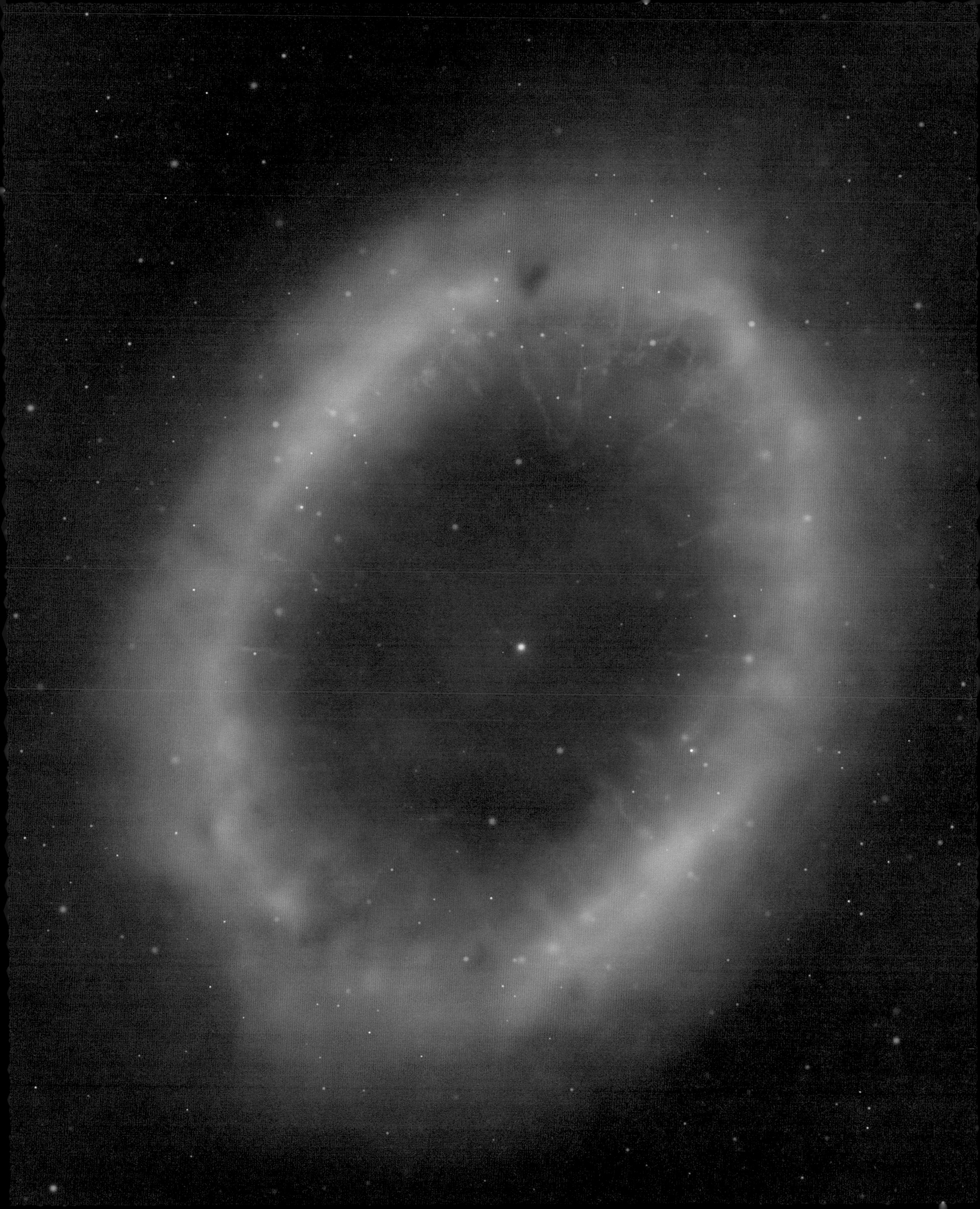

12 365 million years
White Dwarf

Image opposite: The Sun is dead. All that remains is a tiny glowing ember called a white dwarf. At first the star will be fairly luminous by virtue of its temperature. But as it cools down it will grow steadily dimmer. This is a view from the Earth when the white dwarf has cooled to the extent that it provides barely more light than a few full moons.

Image above: The white dwarf Sun has only 50 per cent of the mass it was born with. This makes it only about 1.5 times the size of the Earth.

The exposed stellar core is also a star in some sense – but it is by no means an ordinary star. The almost continuous contraction that it has suffered since it left the main sequence, unsupported by nuclear reactions, has compressed it to the extent that it is now only 1.5 times the diameter of the planet Earth. At this size, the bizarre object has an almost unimaginable density. In that tiny volume resides enough material to make up just over half the Sun: half a tonne squeezed into every cubic centimetre. By now the star's nuclear reactions have fizzled out. But they have left their mark. Over billions of years they have converted what was originally hydrogen into helium, carbon and oxygen. And their ferocious heat has left the star with an exceedingly high surface temperature of about 120 000 Celsius. At this temperature, the star, despite its puny surface area, is quite luminous, about 35 times brighter than the Sun as it shines today.

This, then, is the Sun's legacy: a white hot, compact stellar corpse only 1 per cent of its main-sequence size, composed mainly of carbon and oxygen. Astronomers call such an object a white dwarf.

White dwarfs do shine of their own accord. Yet they are not normal stars in the sense that they no longer support themselves through nuclear reactions or even produce new energy. At such densities, these stars do not need nucleosynthesis to hold their own in the tug-of-war against gravity. Quite simply, white dwarf matter is so closely packed that it almost cannot physically be any denser. Such a configuration is called degeneracy. Not even the gravity of half a solar mass of material is enough to squeeze degenerate matter to greater extremes, because subatomic particles – electrons – within the material respond with a so-called degeneracy pressure. This opposes gravity and so keeps the material at the same density. This extreme compactness has another effect as well. It means that the heat generated within the core while it was still a fusion reactor remains locked up inside – and will take a phenomenal amount of time to leak to the surface and escape into space. Thus the white dwarf phase is characterised by a long period of cooling. What was once a yellow star, vibrant with energy, has become little more than a lonely ember in the impossible vastness of space.

Hundreds of billions of years
Black Dwarf

Image opposite: A black dwarf's fate is the same as a white dwarf's: a steady cooling down. Gradually the object disappears from the infrared part of the spectrum, the microwave, then the radio region – until it emits almost no detectable form of radiation whatsoever. In the end, perhaps 100 billion years after it formed, it betrays it presence only in silhouette, by the light it obscures from stars and nebulae behind it.

After about 10 billion years, the surface of the white dwarf Sun cools to around 3000–4000 Celsius. At these temperatures the object looks distinctly red (even though it is still called a white dwarf), and is tens of thousands of times dimmer than the main-sequence star it used to be. By now the cooling rate, which slows drastically with age, is incredibly slow. But though it takes an unfathomable amount of time – longer even than the current age of the Universe – the white dwarf Sun one day vanishes totally from the optical window in the electromagnetic spectrum through which we humans today admire the Universe. Too cold to emit any signs of optical radiation at all, the dead Sun ceases to shine. It becomes a black dwarf.

At long last, perhaps 100 billion years from now, maybe even longer, the light will go out in the Solar System. The battered planets still remain, their orbits being stable, huddled around a star that they can no longer 'see'. But overall, the scale of the planetary realm is almost twice as large as it is today. At only half a solar mass, the dark star's gravity clings quite feebly to its retinue of charred worlds, and they each orbit about 1.85 times further out than they do today. Meanwhile, those planets look little, if anything, like the worlds we know today. Over the tens of billions of years, facing a steadily declining heat source, the terrestrial planets have cooled down to just a few degrees above the coldest temperature possible, absolute zero. Their atmospheres have long since vanished, swept away when the Sun was a red giant star so many billion of years earlier. Their surfaces are cracked and battered, free of all ices – and utterly hostile. And in the face of such punishingly cold and dark conditions, even the once majestic giant planets, from Jupiter to Neptune, have frozen nearly solid, contracting slightly as they did so. Now they resemble gigantic spheres of hydrogen-rich ice, petrified and cracked during their slow contraction, their rings stripped away long ago.

Some asteroids, but not many, remain also, orbiting in the eternal night. But the content and size of the Oort cloud are substantially different this deep into the future. Many comets will have been wrenched away from the cooling white dwarf during the frequent encounters with nearby, passing stars. And with the white dwarf having a smaller mass and gravity than the Sun today, other comets have escaped from the dead Sun completely. They will spend eternity drifting between the stars – if they don't get captured and integrated into Oort clouds around other stars.

The Solar System as we know it is dead. But, from the gases discarded by the dying Sun, something wonderful will one day spring.

Time unknown
End of an Era ... Start of an Era

Image opposite: Something old, something new. In a time too far from now to comprehend, a massive star reaches the end of the line and blows itself apart in a supernova. Shockwaves from this event will compress nearby gases – some of them once part of a yellow star called the Sun – and stimulate the gravitational contraction that will turn them into new stars.

Time goes by. And the dead Solar System continues to orbit the galactic centre just as it does today. Its journey takes it past stars young and old; carries it through the ghostly ruins of stars already dead – shrouds of gas ejected by planetary nebulae, or during the cataclysmic explosions known as supernovae. Ultimately the ashes of the star once known as the Sun are spread throughout the entire Milky Way galaxy, mixed indistinguishably with the remains of other stars, replenishing the interstellar medium from which they sprang billions of years earlier.

Then one day...

A massive star reaches the end of its life. It blows itself apart. Shockwaves from the supernova spread outwards from the epicentre through the interstellar medium like concentric rings on the surface of a lake. The waves compress the gas clouds through which they propagate. And eventually, somewhere, part of the cloud begins to contract under its own gravity. Millions of years later, a new star shines in the galaxy, born from the ashes of those long dead – including the Sun. Perhaps planets will also form around this new star – even life. And so it could be that the very atoms that currently comprise our bodies will one day find themselves part of a very different, alien creature. For the Universe is the ultimate recycling machine.

We have come full circle.

Glossary

accretion
The process through which a body gradually increases its mass by gravitationally pulling in smaller masses that then stick to it. The planets grew by accretion in a protoplanetary disc.

agglomeration
The process whereby grains in the Solar Nebula grew larger by jostling up alongside and sticking to their neighbours, either by adhesive or electrostatic forces. It differs from accretion in that gravity plays almost no role because the particles involved are too small.

anorthosite
The stuff from which most Moon rocks are made, a calcium–aluminium–oxygen compound.

aphelion
The furthest point from the Sun in the orbit of a planet, comet or asteroid.

Ariel
One of the five largest natural satellites of the ice giant Uranus. Ariel is about 580 kilometres across.

asteroid
A heavily cratered stony or metallic body found throughout the Solar System, but especially in the asteroid belt between Mars and Jupiter. The smallest asteroids so far detected are a few metres across, though they range in size from dust grains upwards. The largest asteroids, such as Ceres and Vesta, are much more massive, hundreds of kilometres across. All of them are the debris from the planet-formation process, and are among the oldest objects in the Solar System.

asteroid belt
The band between the orbits of Mars and Jupiter where most of the Solar System's asteroids reside. The belt extends from about 2 AU to 3.3 AU from the Sun, and at these boundaries the asteroids within the belt take about three years and six years, respectively, to orbit the Sun.

asthenosphere
The top part of the Earth's mantle, about 400 kilometres beneath the surface, where most of the material is molten.

astronomical unit (AU)
The distance from the Earth to the Sun, roughly 150 000 000 kilometres.

atom
The smallest unit to which a chemical element can be reduced and still recognised as that element. Atoms have dense, positively charged nuclei made of protons and neutrons, orbited by negative electrons. The electron count is the same as the proton count, making the atom overall electrically neutral.

basin
A particularly large impact feature formed by asteroids or comets, usually several hundred kilometres across and often flooded with solidified lava.

black dwarf
A white dwarf that has cooled down so substantially, over many billions of years, that it no longer emits optical light. No black dwarfs have ever been found – the Universe is not yet old enough to have produced any.

Callisto
The innermost of Jupiter's four Galilean satellites and the second largest. Callisto's icy surface is completely covered in impact craters that have accumulated over billions of years without eradication or modification.

Cassini division
The gap, roughly 4700 kilometres wide, that separates Saturn's rings A and B. It was discovered in the seventeenth century by the Italian-born French astronomer Giovanni Cassini (1625–1712).

Ceres
The largest known asteroid, with a diameter of around 900 kilometres, having roughly a third of the combined mass of all asteroids.

Charon
Pluto's only satellite, discovered in 1978. Charon is so large compared with Pluto – one-tenth of its mass and half its diameter – that the pair are considered a double planet.

chromosphere
The name for the innermost part of the Sun's atmosphere, just above the photosphere. The chromosphere can only be seen when the rest of the Sun's glare is masked.

coma
The head of a comet, a ball of dust and ionised gas up to 1 million kilometres across. It shines by light reflected from the Sun.

comet
An icy body in orbit around the Sun. Most comets are found in a vast reservoir known as the Oort cloud, tens of thousands of AU from the Sun, where they are inert and frozen. When they venture into the inner Solar System, however, the Sun's heat melts their surface ices, which then escape into space to form the coma and tail that makes these objects so spectacular. Comets may be just a few tens of kilometres across, but their tails can extend for more than 100 million kilometres.

condensation
The process whereby a substance, on cooling down, turns from a vapour to a liquid or solid. Condensation was the mechanism that started to build the first solid grains in the Solar Nebula.

convective zone
The outermost 30 per cent of the Sun. In the convective zone, heated gas rises in discrete packets, called convection cells, until it reaches the surface. Once there, the convective cells spread out, their cargoes of gas cool down, and the gas then sinks again, later to be reheated. The individual cells are visible on the Sun's surface as the solar granulation.

core
The central 30 per cent of the Sun where temperatures are high enough – at 15 million Celsius – to generate the Sun's nuclear reactions. In the core, hydrogen is slowly converted into helium. The pressure this generates prevents the Sun from collapsing under gravity and will keep it shining for several billion years. The term is also used to describe the dense inner parts of the planets.

corona
The outer atmosphere of the Sun. The corona is very hot at 2 million degrees Celsius and subsequently highly ionised. It is the source of the Sun's stellar wind.

crater
An impact crater is a circular or elliptical depression formed when a solid planetary body is struck at speeds of several kilometres per second by an asteroid, comet or other fast-moving projectile. Volcanic craters, which are less circular, are formed when terrain collapses above a magma reservoir that has emptied through constant eruption. They can also be created during volcanic explosions.

crust
The outermost part of a terrestrial planet or satellite, extending all the way to the surface. The terrestrial planets have rocky crusts. But the moons of the outer planets have crusts that are rich in ices.

Dactyl
The tiny moon of asteroid 243 Ida. Dactyl is just 1 kilometre long.

Deimos
The outermost of the two satellites of Mars. Deimos is an irregular lump of rock, with dimensions of roughly 15 × 12 × 11 kilometres. It is very likely an asteroid or planetesimal captured by Mars in the past.

Dione
A mid-sized moon of Saturn, composed of ice and rock and about 1120 kilometres in diameter.

dust
In astronomical terms, tiny particles, usually less than one-millionth of a metre across, in the interstellar medium. Interstellar dust grains are elongated, and are made chiefly of graphite, silicates and iron. Ices may also be present.

dwarf
A star converting hydrogen to helium in its core, and thus on the so-called main sequence. Dwarfs are generally smaller than those stars that have begun to run low on, or have exhausted, their hydrogen supply – hence the name. The Sun is a yellow dwarf star.

Earth
The third planet from the Sun – at a distance defined as 1 AU. Earth is the largest of the four terrestrial planets and the only one capable of sustaining liquid water at the present epoch. It is one of only three volcanically active bodies in the known Solar System. This activity, and the substantial wind erosion and water erosion, mean that Earth's surface is geologically young, with few fresh impact craters. The atmosphere, which is substantial, is composed of 77 per cent nitrogen, 21 per cent oxygen – a result of the planet's diverse plant life – 1 per cent water vapour and 0.9 per cent argon. An ozone layer shields the surface from harmful ultraviolet radiation. Earth also has a substantial magnetic field, second only in strength to that of Jupiter.

eccentricity
A number representing the degree of ellipticity, or deviation from circularity, of an orbit. An eccentricity of zero means a circular orbit. The planets have eccentricities between zero (Venus and Neptune) and 0.25 (Pluto), while comets and some asteroids have much larger eccentricities.

ecliptic
The plane in which the Earth orbits the Sun. Most planets orbit close to the ecliptic also, with the exception of Mercury and Pluto.

electron
A negatively charged subatomic particle usually found orbiting an atomic nucleus. Electrons can roam freely when they are removed by intense radiation (ionisation), and are known as negative ions.

ellipse
A closed curve symmetrical about two axes that looks like a squashed circle. An ellipse is the path followed by a planet around the Sun according to the laws of German astronomer Johannes Kepler. The Sun is not at the centre of the ellipse, but off to one side at one of two points called foci (singular, focus).

Enceladus
A small ice moon of Saturn, about 500 kilometres across.

Eros
Asteroid number 433, measuring 36 × 15 × 13 kilometres. Eros made history when it became the first asteroid to be orbited by a probe – NEAR-Shoemaker in 2000.

escape velocity
The speed needed to escape completely from the gravity of a planet or other large body. For example, a projectile must reach a speed of 11.2 kilometres per second if it is to escape from the Earth.

Europa
The smallest of the four Galilean satellites of Jupiter, and the second furthest of them from that planet. Europa's surface is a solid shell of water ice, cracked by internal geological activity. Planetary geologists speculate that an ocean of liquid water might exist under the icy exterior, which may only be a few kilometres thick.

extrasolar planet (exoplanet)
Any planet orbiting a star other than the Sun. Most of those so far found are giants larger than Jupiter. This is not because no smaller ones exist, but because current technologies and detection methods are only sensitive to very massive planets.

fault
A geological feature associated with tectonic activity. Faults are cracks in a planet's crust brought about by lateral surface movements and stretching.

Galatea
A small moon of Neptune, just 160 kilometres across, which orbits just inside the outermost ring, Adams. Galatea's presence herds together the particles in Adams and creates the ring's arcs.

Galilean satellites
The four largest moons of Jupiter: Io, Europa, Ganymede and Callisto. They are named after their discover, Galileo Galilei (1564–1642).

Ganymede
The largest planetary satellite in the Solar System, bigger even than the planet Mercury and half as large again as Earth's Moon. Ganymede is one of the four Galilean moons of Jupiter, the third one from that planet in terms of distance. About two-thirds of its icy surface is in the form of a light terrain that has many grooves but relatively few craters. The remaining third is dark and heavily cratered.

gas giant
The name applied to the two largest planets in the Solar System: Jupiter and Saturn. Gas giants are composed chiefly of elements that on Earth we know as gases – hydrogen and helium for example. However, the term gas giant is a misnomer in the sense that these elements exist on these planets mainly in liquid rather than gaseous form, because of the high pressures found inside planets.

Gaspra
Asteroid number 951, measuring 17 kilometres long. In 1991, Gaspra was photographed by the Galileo spacecraft.

globule
A dense, dark and roughly spherical nebula inside which a star or many stars are forming. Globules are stellar cocoons.

granulation
The patchwork pattern on the surface of the Sun, which derives from convective transfer of heat from the interior to the surface.

Great Dark Spot
A dark blue patch in the atmosphere of Neptune. The blemish, a hurricane-like storm similar to, though smaller than, Jupiter's Great Red Spot, was discovered in 1989 when the Voyager 2 spacecraft flew past Neptune. It measured about 5000 kilometres deep by 10 000 kilometres across.

Great Red Spot
A large cyclonic storm feature in Jupiter's atmosphere, measuring about 14 000 kilometres by up to 40 000 kilometres. The Great Red Spot's colour may derive from the presence of compounds such as phosphine.

greenhouse effect
An increase in the global temperature of a planet as a result of certain constituents in the planet's atmosphere. On Earth, for example, sunlight (visible radiation) filters through the atmosphere and is absorbed by the surface and re-emitted at infrared wavelengths. This radiation cannot escape back into space because carbon dioxide, methane and other components in the atmosphere are opaque to infrared energy. As a result the planet's heat is trapped, making it hotter than it would be if these so-called 'greenhouse gases' were absent. On Venus, the greenhouse effect has reached extremes because of the large quantity of carbon dioxide in Venus' atmosphere, and the planet is 200 Celsius hotter than it would otherwise be.

heavy bombardment
The period right after the formation of the planets and satellites during which their surfaces were bombarded by leftover, unaccreted planetesimals and comets. The bombardment phase lasted for about 1200 million years, from 4500 to 3300 million years ago.

helium
The second most abundant element in the Universe, after hydrogen, and the second lightest. Helium is found in large quantities in the Sun (and other stars) and in the giant planets. Its nucleus is composed of two protons and two neutrons.

helium flash
The sudden ignition of helium burning that occurs in the core of a red giant after the core has been sufficiently compressed and its temperature reaches 100 million Celsius. The ignition is explosive and occurs within mere seconds of attainment of the correct conditions – hence the term 'flash'.

Herbig–Haro object (HH object)
A bright nebula usually associated with starbirth. Herbig–Haro objects – named after the astronomers George Howard Herbig and Guillermo Haro – are created when jets from young stellar objects plough into nearby interstellar gas and make it glow.

hydrogen
The most abundant of all the elements and the lightest and simplest. Hydrogen is found throughout the Universe – in giant molecular clouds, in smaller nebulae, in stars and in giant planets such as Jupiter. Its nucleus is a single proton. Hydrogen is the fuel that drives stars, converted into helium in their cores.

Iapetus
The third largest moon of Saturn, after Titan and Rhea. It has a curious coating of dark compounds that covers an entire hemisphere, starkly contrasting with the other bright and icy hemisphere.

Icarus
Asteroid number 1566, famous because its highly eccentric orbit brings it even closer to the Sun than Mercury.

ice giant
The term applied to the two giant planets Uranus and Neptune. Compared with the gas giants Jupiter and Saturn, Uranus and Neptune have proportionately higher quantities of icy substances such as methane, ammonia and water ice.

Ida
Asteroid number 243. Ida is fairly large, at 58 kilometres across, and is one of several asteroids known to have a moon, called Dactyl.

interstellar medium (ISM)
The material between the stars, composed of 99 per cent gas – mainly hydrogen and helium – and 1 per cent dust. The ISM is the raw material from which stars are born.

Io
The innermost of the four Galilean satellites of Jupiter, and the most volcanically active body in the known Solar System. Jupiter's gravity constantly flexes Io and this generates the internal heat that drives the volcanoes.

ion, ionisation
An ion is an atom that has lost one or more electrons, making it positively charged. A free electron is also classed as a negative ion. Ions are created when energetic 'particles' of light called photons knock electrons off atoms in a process called ionisation.

jet
Also known as a molecular outflow, a stream of particles directed away from the central star in a protoplanetary disc in a direction perpendicular to the disc's surface. Jets from young stars such as T-Tauri stars are common. Those that plough into surrounding interstellar matter cause it to glow conspicuously and this leads to the formation of Herbig–Haro objects.

Jupiter
The fifth planet from the Sun and the largest in the Solar System. Jupiter's mass is more than double that of all the other planets combined. Jupiter is a gas giant, and is composed chiefly of hydrogen and helium. However, at the extreme pressures found inside Jupiter, most of this material is in a liquid state. A solid core of rock and ice, perhaps 15 times as massive as Earth, may also be present. Because of its rapid rotation – once every 9.8 hours – Jupiter is flattened at the poles, and the coloured clouds that reside in its atmosphere are stretched into bands that completely encircle the planet.

Kirkwood gaps
Apparent gaps in the asteroid belt where there are fewer asteroids than normal, the direct result of gravitational perturbations induced by the planet Jupiter. They are named after the American astronomer who brought attention to them, Daniel Kirkwood (1814–95).

Kuiper belt
A donut-shaped reservoir around the Sun, composed of at least 10 million dormant, icy comet nuclei. The Kuiper belt is constrained largely to the plane of the Solar System, and extends from about 30 AU – roughly the position of Neptune – to perhaps 30 times that distance. Some astronomers consider the Kuiper belt an inner extension of the much more distant and spherically shaped Oort cloud.

Kuiper-belt object (KBO)
A cometary body orbiting in the Kuiper belt, beyond Neptune. Also known as a trans-Neptunian object (TNO).

lithosphere
On Earth, the transition layer between the solid crust and the liquid mantle or asthenosphere. The lithosphere is rigid, unlike the asthenosphere, and is divided into continental plates. It is 80 kilometres deep below the oceans and about twice that below dry land.

main sequence
The longest part of a star's life span, during which it converts hydrogen to helium in its core and is known as a dwarf. The Sun is currently on the main sequence.

mantle
The part of the interior of a satellite or terrestrial planet that lies between its core and its crust. The mantle is usually solid, but on Earth it is partially molten nearer to the surface, and even the solid parts flow like a liquid because of their extreme environments.

maria (singular, mare)
The Latin word meaning 'seas' used to describe the smooth, dark plains seen on the Moon. Maria are solidified lakes of basalt lava, which flooded the surface when the Moon suffered heavy impacts.

Mars
The fourth planet from the Sun and second smallest of the terrestrial planets, after Mercury. Mars was once a volcanically active place, and evidence is in abundance that liquid water once flowed on its surface. Now, however, the planet appears dead. Mars has a tenuous atmosphere of carbon dioxide, 100 times thinner than Earth's atmosphere. Its two irregular moons, Phobos and Deimos, are captured planetesimals.

Mercury
The closest planet to the Sun and second smallest, after Pluto. Mercury's surface is highly cratered and cracked – the cracks perhaps caused by slow contraction – and in many ways it resembles the Moon's. Aside from Pluto, Mercury's orbit is the most eccentric in the Solar System. Mercury is also the second densest planet, its iron core taking up much of its interior.

meteorite
A body from space substantial enough to survive the heat of entry through an atmosphere and land on a planet's surface. Most meteorites are fragments of asteroids. Meteors, by contrast, are smaller and never reach the surface because they burn up.

Mimas
A mid-sized, icy satellite of Saturn, about 400 kilometres across. Mimas is dominated by a large crater almost half the size of the moon itself, giving it a 'Death-star' appearance.

minor planet
See asteroid.

Miranda
The fifth largest of the five classical satellites of Uranus, just 480 kilometres across and not quite spherical. Miranda is famed for cliffs that jut several kilometres above the local surface.

molecular cloud
A vast assemblage of gas and dust particles from which stars form. Molecular clouds range in size up to several hundred light-years, and are composed chiefly (73 per cent by mass) of molecular hydrogen gas – in which the smallest particles are molecules of two hydrogen atoms. Helium and dust account for 25 per cent and 1 per cent by mass respectively. Temperatures in molecular clouds are very low, less than −250 Celsius, and average densities are only a few thousand molecules per cubic centimetre.

molecular outflow
See jet.

Moon
The natural satellite of Earth. The Moon's surface is heavily cratered, and great areas have been flooded in the past by seas – maria – of liquid basalt that were brought to the surface by meteoritic impacts or volcanism and have since solidified. The Moon's composition suggests it was formed partly from the Earth and partly from another, unknown body.

moon
See satellite.

near-Earth object (NEO)
An asteroid whose orbit crosses that of the Earth.

nebula
A cloud of gas and sometimes dust in space. Some nebulae are bright, glowing because of the energising radiation of the stars within them. Others shine only by the light they reflect from nearby stars. And still others, those with a significant dust content, are dark, visible only by the background stars and nebulosity that they obscure.

Neptune
The eighth planet from the Sun and the second of the two ice giants. Neptune is about four times larger than Earth – slightly smaller than Uranus – and has a blue hydrogen–helium atmosphere. Like Uranus, Neptune is believed to harbour a rock and ice core roughly the size of the Earth. This is surrounded by a mantle of hydrogen-based ices such as methane, water and ammonia. Neptune has one large irregular satellite, Triton, several smaller ones, and a dark system of rings.

Nereid
A 340-kilometre irregular satellite of Neptune with a high inclined, distant and elongated orbit.

nucleus
The positively charged core of an atom. The central part of a comet is also called the nucleus.

Oberon
The outermost of the five icy, classical satellites of Uranus, its surface heavily cratered. Oberon is the second largest Uranian moon after Titania, and is about 1520 kilometres across.

Olympus Mons
The largest known volcano in the Solar System, on Mars, now probably extinct. Olympus Mons stands 27 kilometres above the local ground level and is about 600 kilometres in diameter.

Oort cloud
An extensive reservoir around the Sun that is home to trillions of inert comet nuclei. The Oort cloud is roughly spherical and may extend as far from the Sun as a light-year – a significant part of the way to the nearest stars.

orbit
The path of a planet around a star or a satellite around a planet. Most planets and satellites follow slightly elliptical (almost circular) orbits, according to the laws of German astronomer Johannes Kepler, whereas comets and asteroids have elliptical orbits that are much more elongated.

outgassing
The process by which gases, chemically bonded to the minerals in rocks, are released as the rocks are heated and the bonds dissolved. The current atmospheres of Mars, Venus and Earth were created in part by outgassing, comets bringing many of the other constituents.

perihelion
The closest point to the Sun in the orbit of a planet, comet or asteroid.

Phobos
The innermost of the two Martian satellites, and very likely a captured asteroid or planetesimal like its partner, Deimos. Phobos is irregularly shaped and measures 27 × 22 × 19 kilometres.

photon
A 'particle' of electromagnetic radiation, of which light and radio waves are examples.

photosphere
The 'surface' of the Sun defined as the radius out from the Sun's core where the Sun's gases finally become opaque.

planet
Generally, a large non-luminous body in orbit around a star, shining only by the light it reflects, made of either gaseous elements or rocky and metal substances. Planets are not the only things that can orbit a star, however. Comets and asteroids are found in great numbers in the Solar System, and are distinguished from planets by their relatively small sizes. However, the size a body has to be before it is classed as a planet is not well defined.

planetary nebula
The outermost layers of a dying star, ejected into space when the star begins to run out of fuel in its core and becomes unstable. It is estimated that 95 per cent of all stars, including the Sun, are destined to become planetary nebulae. They last a few tens of thousands of years, over which time they grow to several tenths of a light-year across. But, as they grow, they become thinner and fainter, and gradually blend into the surrounding interstellar gas and vanish. In so doing they recycle the Universe's material, and so pave the way for future generations of stars. Planetary nebulae have nothing to do with planets, and are so named merely for historical reasons.

planetesimal
A body of metal, rock and ice presumed to have formed in the early stages of the Solar System by accretion of smaller fragments. Planetesimals are about 0.1–100 kilometres across. Today, millions of rocky and metallic planetesimals populate the asteroid belt. The icy planetesimals, which formed beyond the snow line, survive as comets in the Kuiper belt and Oort cloud.

plasma
A gas whose atoms have been ionised, their electrons knocked off by intense radiation. The Sun is made of ionised hydrogen and helium, so it is essentially a ball of plasma.

plutino
The name often given to those Kuiper-belt bodies beyond Neptune whose orbits are in a 3 : 2 resonance with that planet – they orbit twice for every three orbits of Neptune. Pluto is in a similar resonance orbit, hence the name plutino.

Pluto
The smallest planet in the Solar System – smaller even than our Moon. Pluto, which has one natural satellite, called Charon, is usually the furthest planet from the Sun. But its highly elliptical and therefore unusual orbit – which extends from 30–50 AU – means than it occasionally passes inside the orbit of Neptune. The planet is very lightweight, but massive enough to have a rocky core. This is covered with ices of water, ammonia, methane and nitrogen. The surface temperature is so cold, at around −230 Celsius, that Pluto is able to hold onto a thin atmosphere of nitrogen and methane. These gases freeze out, though, when the planet's elliptical orbit takes it far from the Sun.

Proteus
The second largest satellite of Neptune, but much smaller than Triton at only 400 kilometres across. It is an irregularly shaped lump of rock and ice.

proton
A subatomic particle having the same charge as an electron but being positive rather than negative. They usually combine with neutrons to form atomic nuclei, in turn orbited by electrons. The only element whose atoms have no neutron is hydrogen.

protoplanet
A primitive planet formed early in the Solar System from the lumping together of planetesimals. As the protoplanets orbited the Sun, they gradually swept up the planetesimals in their path through the process of accretion, and grew into the planets we see today.

protoplanetary disc
A disc of gas and dust surrounding a star, inside of which planets are forming – or will shortly begin forming.

protostar
A cool object, comparable in size to the orbits of the inner planets, which is contracting and will one day form a new star. A protostar is not yet dense enough, nor its interior hot enough, to drive the thermonuclear reacions that drive true stars.

radiation pressure
A force exerted by light, just like a hail of bullets. It arises because radiation is made up of tiny packets of energy called photons.

radiative zone
The region outside the Sun's core but before the convective layer. Energy generated in the Sun's core travels through the radiative zone via radiation or streams of particles called photons.

red giant
A star that has run out of hydrogen gas in its core and whose outer layers have expanded enormously and cooled in response to the shrinking core. The Sun will turn into a red giant in several billion years.

regolith
The powdery substance that covers the surface of the Moon and other planetary bodies, the remains of surface rocks that have been pulverised by billions of years of meteoritic impacts.

Rhea
The second largest moon of Saturn. At only 1500 kilometres across, Rhea is a lot smaller than Titan, the largest Saturnian satellite.

ring arcs
Arc-shaped segments in a planetary ring – notably Neptune's Adams ring – where there is an overabundance of ring fragments. They are caused by the gravitational effects of nearby moons.

ring system
A collection of particles surrounding a planet and confined to orbit it in an extremely thin plane. Only the four giant planets have rings. Each ring system is unique in brightness, size, and in the size distributions of its constituent particles. Planetary rings may be the result of comets or moons that got torn apart by tidal forces when they approached the planets too closely.

Roche limit
The distance from a planetary body within which any other body, held together mainly by gravity, will be torn apart by tidal forces.

satellite
Any object that orbits a planet. The Moon is a natural satellite of the Earth, whereas the Hubble Space Telescope is an artificial one.

satellite, irregular
Those natural satellites not formed in discs around their planets, but rather gravitationally captured later. Irregular satellites are usually small and non-spherical. They do not all orbit in the same plane, nor even always in the same direction.

satellite, regular
Any natural satellite that forms in a disc around a newly forming giant planet. As a result of their formation, the regular satellites orbit their parent planets in a common plane and each in the same direction as the others.

satellite, shepherd
A natural satellite whose gravitational effects act either to keep planetary rings narrow, or to sharpen their edges.

Saturn
The second largest planet in the Solar System, after Jupiter, and sixth from the Sun. Like Jupiter, Saturn is a gas giant and has a similar construction: hydrogen and helium surrounding a solid core of rock and ice. Saturn is most famous for its rings, which are the brightest of all ring systems.

scarp
A surface fracture found on some planets and satellites. Scarps are created when surface crust contracts as it cools, or moves sideways as a result of tectonic activity.

shell burning
A phase in a star's life when its reactions take place not in the core – which has become saturated with non-reactionable material – but in a thin shell on the core's outskirts. The Sun will experience shell burning when it becomes a subgiant and a red giant.

snow line
The distance from the Sun beyond which the temperature in the Solar Nebula was low enough to permit the condensation of icy substances. It corresponds roughly to the orbit of Jupiter.

Solar Nebula
The vast pancake of gas and dust that surrounded the newly forming Sun, from which the planets, asteroids and comets later formed. The Solar Nebula measured 100–200 AU in diameter. Similar so-called protoplanetary discs are now seen around nearby stars.

Solar System
The name given to the system of planets, satellites, asteroids and comets that surrounds the Sun. The Solar System has nine major planets. The first four (Mercury, Venus, Earth and Mars) are primarily made of rock and metal – they are the terrestrial planets. The next two (Jupiter and Saturn), which are very much larger, are made mostly of hydrogen and helium and are known as gas giants. The next two (Uranus and Neptune) are intermediate in mass. They are also made chiefly of hydrogen and helium. But, because most of the hydrogen is locked up inside ice molecules such as methane or water, Uranus and Neptune are referred to as the ice giants. All of these planets except Mercury and Venus have natural satellites. The last planet, Pluto, is an oddball smaller even than some of the planetary satellites. Aside from the planets, the Solar System sports a wide belt of asteroids, and two zones of comets are found beyond the orbit of Neptune. Often the term 'solar system' (lower case) is use to refer to planetary systems other than the Sun's.

solar wind
The steady sea of charged particles – electrons, protons and some atomic nuclei – that flows away from the Sun's corona. The solar wind grows naturally thinner with distance from the Sun, and can be considered to end at about 100 AU – roughly two times the maximum distance of Pluto.

star
A self-luminous body, that for most of its life shines by the energy it generates through the fusion reactions in its innermost region, the core.

stellar wind
A stream of charged particles flowing away from a star, similar to the Sun's own solar wind. Some stars have very massive, fast winds, such as the young T-Tauri stars.

subgiant
A star in which the core hydrogen has been consumed, and nuclear reactions take place only on the outskirts of the core in a thin shell. The Sun will become a subgiant one day, and as a result will swell to about two or three times its current diameter.

sublimation
The process of turning directly from a solid to a gas without passing through a liquid phase.

subsolar point
The point on a planet's or moon's surface where the Sun is directly overhead.

Sun
The star at the centre of the Solar System. The Sun is a yellow dwarf with a surface temperature of about 5700 Celsius and a core temperature of some 15 million Celsius. It is currently about 4600 million years into its main-sequence lifetime and will remain on

the main sequence for several billion years more before its hydrogen supply runs out and it begins to die.

sunspot
A region of enhanced activity on the Sun. Stronger magnetic forces hinder the flow of surface gases and make these regions 500–1600 cooler than average. From a distance these zones show up as dark sunspots.

supernova
A cataclysmic explosion at the end of a massive star's life. Like planetary nebulae, supernovae explosions replenish the interstellar medium and supply the materials for future generations of stars. The Sun will not become a supernova because it is not massive enough.

tectonics
Lateral movements in the crust of a planet or satellite that lead to crustal deformation landforms. Tectonic features take the form of stretch marks, faults and fractures called scarps.

terrestrial planet
One of the four innermost planets in the Solar System: in order of size, smallest first, they are Mercury, Mars, Venus and Earth. These planets are small – diameters of 12 756 kilometres or less – and composed chiefly of rocky and metallic materials. Mercury is the only one to lack an atmosphere.

Tethys
A mid-sized, icy moon of Saturn, 1060 kilometres in diameter.

tide
A force exerted on one planetary body by another by virtue of the diminishing effects of gravity with distance. The Moon, for example, exerts a gravitation pull on the face of the Earth nearest to it, but it exerts a slightly smaller pull on the other face, further away. The difference between these two pulls is the tidal force.

Titan
The largest satellite of Saturn and the second-largest satellite in the Solar System after Jupiter's Ganymede. Titan, half rock and half ice, is the only satellite with a substantial atmosphere, made of nitrogen. Currently the surface geology is unknown.

Titania
The largest of the five classical satellites of Uranus, about 1580 kilometres across.

trans-Neptunian object (TNO)
See Kuiper-belt object.

Triton
The largest satellite of Neptune, about 78 per cent the size of Earth's Moon. Triton is an ice moon with the coldest measured surface in the Solar System at only −235 Celsius. Thus it is somewhat surprising that the body is active: geysers spout fountains of nitrogen gas and organic molecules several kilometres above the moon's surface. There they are caught in the prevailing winds – Triton has a tenuous atmosphere – and blown downwind where they streak and stain the surface. Triton's odd orbit, highly inclined and retrograde, implies that it is an irregular satellite, a captured protoplanet.

T-Tauri star
A young star, less than 10 million years old, not yet on the main sequence. T-Tauri stars are the precursors of Sun-like stars. But they are slightly cooler and redder and are several times larger than their eventual main-sequence diameters. They have fierce stellar winds and often exhibit extremely violent episodes of magnetic activity. The Sun went through a T-Tauri phase several million years into its lifetime.

Umbriel
About 1170 kilometres in diameter, the third-largest of the five classical satellites of Uranus. Its surface is heavily cratered.

Uranus
The seventh planet from the Sun. Uranus is four times larger than Earth and has a greenish-coloured and featureless hydrogen–helium atmosphere. Underneath the atmosphere the planet is thought to have a rock and ice core roughly the size of the Earth, surrounded by a mantle of hydrogen-based ices such as methane, water and ammonia. Uranus spins on its side, and so may have experienced a dramatic collision with another icy protoplanet early in the planet's history. Uranus has five large, classical satellites, several smaller ones and a system of dark rings.

Venus
The second planet from the Sun and second largest of the terrestrial planets. Venus is comparable in size and mass to the Earth and has been called Earth's twin. But a runaway greenhouse effect has raised the temperature on Venus to the point that tin and lead would melt there. A choking atmosphere of carbon dioxide and clouds of concentrated sulphuric acid make Venus the one place in the Solar System most like Hell. The planet has no satellites, and spins backwards compared with most other planets.

Vesta
The fourth asteroid to be discovered and the third largest, measuring some 576 kilometres across.

volcanism
The process whereby magma – molten rock – reaches the surface of a planet through fissures in its solid crust. As the magma piles up on the surface it creates the landforms typical of volcanism: cones, domes and other features. Sometimes the magma contains trapped gas under pressure. When this reaches the surface it does so explosively. Volcanoes are found throughout the Solar System, but only those on the Earth, Io and Triton are definitely still active.

white dwarf
The next stage in the life of a low-mass star, after it has become a red giant and has ejected its atmosphere into space to form a planetary nebula. White dwarfs are initially hot and white – hence the name. But, without an internal heat source, they are destined only to cool down forever. When the white dwarf has cooled to the point that it no longer shines, it is called a black dwarf.

Index

Numbers in bold are page numbers for illustrations.

F

G

H

I

J

K

L

M